Plant Cell Biology
an ultrastructural approach

Brian E. S. Gunning
Professor of Developmental Biology
Australian National University, Canberra

Martin W. Steer
Lecturer in Botany
The Queen's University of Belfast

ne, Russak & Company, Inc.
York

© Brian E. S. Gunning and Martin W. Steer 1975

First published 1975
by Edward Arnold (Publishers) Ltd.

Library of Congress Catalog Card No. 75-13749
ISBN 0-8448-0669-2

Published in the United States by:
Crane, Russak & Company, Inc.
347 Madison Avenue
New York, N.Y. 10017

All Rights Reserved. No part of this publication may be reproduced, stored in a retrieval system, or transmitted in any form or by any means, electronic, mechanical, photocopying, recording or otherwise, without the prior permission of Edward Arnold (Publishers) Limited.

Filmset by Photoprint Plates Ltd, Rayleigh, Essex
Printed in Great Britain by Fletcher & Son Ltd, Norwich

List of Plates

Plate 1   The Plant Cell (1). Light Microscopy
Plate 2   The Plant Cell (2)
Plate 3   The Plant Cell (3)
Plate 4   Plasma Membrane, Microfibrils in the Cell Wall
Plate 5   Xylem (1): Developing Xylem Elements
Plate 6   Xylem (2): Mature Xylem and Xylem Parenchyma
Plate 7   Phloem (1): Sieve Element and Companion Cell
Plate 8   Phloem (2): Sieve Plates and Sieve Pores
Plate 9   Wax and Cuticle
Plate 10   A Capitate Gland
Plate 11   Pollen Grains (1): Developmental Stages
Plate 12   Pollen Grains (2): The Mature Wall
Plate 13   Transfer Cells
Plate 14   Plasmodesmata
Plate 15   Pits
Plate 16   Endodermis and Casparian Strip
Plate 17   Vacuoles
Plate 18   The Nuclear Envelope and its Pores
Plate 19   The Nucleolus
Plate 20   The Endoplasmic Reticulum, Polyribosomes, and Protein Synthesis in Cotyledon Cells
Plate 21   The Endoplasmic Reticulum and Polyribosomes
Plate 22   The Cytoplasm of Tapetal Cells
Plate 23   Smooth Endoplasmic Reticulum in 'Farina' Glands
Plate 24   Developmental Changes in the Endoplasmic Reticulum of Sieve Elements
Plate 25   The Membranes of Dictyosomes
Plate 26   Production of Scales in the Golgi Apparatus
Plate 27   Relationships between Dictyosomes, Endoplasmic Reticulum, and Nuclear Envelope
Plate 28   The Golgi Apparatus and Mucilage Secretion by Root Cap Cells
Plate 29   Mitochondria (1)
Plate 30   Mitochondria (2)
Plate 31   Plastids I: Proplastids and their Development to Etioplasts and Chloroplasts
Plate 32   Plastids II: Chloroplasts (1)
Plate 33   Plastids III: Chloroplasts (2): Details of Chloroplast Membranes
Plate 34   Plastids IV: Chloroplasts (3): Dimorphic Chloroplasts in the C-4 Plant, *Zea mais*
Plate 35   Plastids V: Chloroplasts (4): Components of the Stroma
Plate 36   Plastids VI: Etioplasts and Prolamellar Bodies (1)
Plate 37   Plastids VII: Prolamellar Bodies (2)
Plate 38   Plastids VIII: The Greening Process: From Etioplast to Chloroplast
Plate 39   Plastids IX: Amyloplasts
Plate 40   Plastids X: Chromoplasts
Plate 41   Microbodies
Plate 42   Cortical Microtubules
Plate 43   Microtubules and Microfilaments
Plate 44   Cell Division (1): Mitosis in *Haemanthus*
Plate 45   Cell Division (2): Prophase
Plate 46   Cell Division (3): Prometaphase and Metaphase
Plate 47   Cell Division (4): Anaphase—Early Telophase
Plate 48   Cell Division (5): Telophase and Cytokinesis
Plate 49   Structure and Function at Intercellular Level

Introduction

More than half of the human brain is concerned with the reception and processing of visual stimuli. With such an emphasis on the sense of sight it is no wonder we say that seeing is believing. More precisely, seeing aids understanding. The aim of the collection of pictures in this book is to allow the visual approach to be used to study plant cells. By scanning and interpreting images of their *structure*, some insight into their *function* can be obtained.

The history of biology tells how often structure has been used as a guide to function, first in relation to easily observed features of plants and animals, and then repeatedly, as newly invented instruments created opportunities to explore new realms of structure. The appearance of cells and tissues in the light microscope is by now familiar to most students of biology. Some light micrographs are included here for purposes of 'orientation', but most of the pictures are concerned with the *next* level of analysis—that provided through the use of the electron microscope.

The resolving power of the electron microscope is about one thousand times better than that of the light microscope, though the present techniques for preparing specimens, being imperfect, do not allow biologists to make full use of this ability to visualize very small objects. Nevertheless in the past two decades the three dimensional architecture of cells and cell components, and even of some large molecules, has been magnified by electron microscopy into the range of our ordinary senses.

The *transmission electron microscope* produces an image of the specimen by passing a beam of electrons through it. Electromagnetic fields manipulate and focus the beam, and the magnified image can be viewed directly on a fluorescent screen or recorded by black and white photography. Because electrons are easily deflected, or scattered, they are given a path that is as nearly as possible collision-free by evacuating most of the molecules of air from within the body of the instrument. It follows that specimens to be placed in the microscope must (a) be strong enough to stand up to the conditions of high vacuum, and (b) be thin enough to transmit enough electrons to give an image, some electrons having been scattered or stopped, thereby creating dark areas in the final micrograph.

Several procedures have been used in the production of the electron micrographs presented here. Some very thin objects (e.g. fragments of cell wall, Plates 4b, c; 26a, c, d) have been *shadow cast*. They were spread on a support film (a thin film of carbon or plastic, the electron microscopist's equivalent of the glass slide used by light microscopists) and sprayed with atoms of metal from a source placed to one side. All exposed surfaces accumulated a deposit that is dense to electrons, while other areas in sheltered shadows remained uncoated and therefore comparatively lucent to electrons. Just as objects imaged in aerial photographs can be measured and identified from the size and shape of the shadow cast, so it is with the shadow-cast electron microscope specimen.

Plates 4a, 17a, 18a, 25a and 33b, c, exemplify the outcome of a most important variation on the shadow casting procedure. The specimen is frozen very rapidly so as to avoid distorting sub-cellular components. Internal surfaces are then exposed by fracturing the frozen material. The fracture plane tends to follow lines of weakness, i.e. regions where there was not much water to start with, and which therefore have not produced strong ice. Cell membranes provide one such region, and so the exposed surface usually jumps from one expanse of membrane to another. Indeed the fracture usually passes along the *mid plane* of cell membranes, fracturing them to reveal an *internal* surface. Details of the surface topography of the fracture plane are then highlighted by subliming off some ice ('etching' the surface). Finally a replica of the surface is prepared in the form of a very thin layer of plastic and carbon. The replica faithfully follows the surface topography of the fracture plane, and the details can be viewed after it has been shadow cast. The whole operation is described as the *freeze-fracturing*, or *freeze-etching*, procedure. The method gives images that are especially trustworthy in being representations of material that was alive at the moment of freezing, and not altered since.

Other methods for looking at surface topography exist. In the *scanning electron microscope* a finely focused beam of electrons is played over the surface. Electrons that are reflected, or caused to be emitted from the specimen, are collected and form the basis of an image that conveys much three-dimensional information. Examples are shown in Plates 6a, b; 9a, b, d, f; 12; 15b-e.

Most of the electron micrographs in this collection were, however, obtained by the technique that is used more than any other by cell biologists. The cells are first chemically fixed, then embedded, and finally sectioned into slices thin enough to be stained and examined in the electron microscope.

By 'chemically fixed' is meant a process in which the normal dynamic and changing state of the cell components is interrupted by the application of *fixatives*—chemicals such as formaldehyde or glutaraldehyde, which rapidly kill the cell and, by forming chemical bridges, cross-link the constituent molecules into a three-dimensional fabric rigid enough to stand up to the subsequent manipulations. The mode of action of

INTRODUCTION

a fixative can be demonstrated by adding some to a solution of a protein such as serum albumin: given suitable concentrations it is not many seconds before chemical cross-linking transforms the fluid solution into a solid mass. A second fixation step, which also functions in staining the specimen, is usually performed. The specimen is transferred from the first fixative to a solution of osmium tetroxide. This highly reactive substance reacts (to varying degrees) with many substances, and can be a cross-linking agent. Some cell components take up more osmium than others and hence, since osmium atoms are dense to electrons, appear darker in the final image. In other words the osmium tetroxide is a differential stain as well as a fixative.

The fixed specimen is not usually strong enough to be sectioned, so it is next dehydrated and a plastic introduced into the spaces formerly occupied by water. Epoxy resins are the most commonly used plastic embedding agents. Once hardened in an oven they possess the necessary strength for sectioning into extremely thin slices. 'Thin' is a rather weak adjective in this context, and the term *ultra-thin* is usually employed when referring to the range of section thicknesses that is acceptable for conventional transmission electron microscopy. The sections have to be so thin that a 1 millimetre thick piece of tissue could (in theory) be further sliced to yield 10 000–20 000 ultra-thin sections.

Examining the contents of a cell by looking at such thin slices in the electron microscope leads to various difficulties of interpretation. It is hard to translate the two-dimensional image that is obtained into the three-dimensional reality. It is rather like investigating a house, its rooms, its cupboards, and all their contents down to 1 millimetre in size, by examining a two centimetre thick slice of the whole building. Obviously it is desirable to look at many such slices, and they should, if at all possible, be cut in known planes or in sequences from which three dimensional reconstructions (e.g. Plate 30c) can be made.

Also, it is vitally important that a general impression of the architecture of the cells and tissues be gained by examining them with the light microscope prior to plunging into the more confusing world of ultrastructure.

The dimensions of the world of ultrastructure are such that unfamiliar units, namely *micrometres* (symbol μm) and *nanometres* (symbol nm), are required:

1 millimetre (mm) equals 1000 μm
1 μm equals 1000 nm
or 1 nm = 10^{-9} metre; 1 μm = 10^{-6} m; 1 mm = 10^{-3} m.

The true size of objects in a micrograph may be calculated using a simple and very useful rule of thumb. There are 1000 μm in 1 mm, therefore a 1 μm object will appear to be 1 mm in size when the magnification is $\times$ 1000. Scale-marker lines placed on a micrograph (e.g. Plate 1) enable the true dimensions of objects to be estimated at a glance. To place a scale-marker representing 1 μm on any micrograph, simply draw a line as many mm long as there are thousands in the magnification. Precise measurements are equally easily obtained, thus:

$$\frac{\text{size in micrograph (mm)} \times 1000}{\text{magnification}} = \text{true size } (\mu\text{m})$$

$$or \frac{\text{size in micrograph (mm)} \times 1\,000\,000}{\text{magnification}} = \text{true size (nm)}$$

All cells possess certain basic biochemical systems that synthesize carbohydrates, proteins, nucleic acids, and many other types of molecule. All have an outer surface that provides protection by excluding harmful material in the external environment, while at the same time permitting the controlled import and export of other substances. All have a store of information where the hereditary material embodies in a chemical form instructions which guide the cells through the intricacies of their development and reproduction. All have devices which provide chemical energy, to be utilized in the general maintenance of cellular integrity, and in syntheses leading to growth and development. These attributes are fundamental to all living systems, and the structural and functional similarities of plant and animal cells stem from them. There are in addition cellular features in which the two kingdoms differ, mostly deriving from one major event in the evolution of living organisms—the development of a cell wall in the ancestors of plants. The consequences for plant cell structure and function were far-reaching.

Plant cells vary in the extent to which different functions are developed, for, as with most other multicellular organisms, plants exhibit division of labour. Specialized cells develop, related to the varied requirements of maintaining life and supporting growth and development—protection, mechanical support, synthesis or storage of food reserves, absorption, secretion, reproduction, cell division, and the humbler but vital role of connecting the more exotic tissues.

Unlike animals, plants tend to restrict processes of cell multiplication to permanently embryonic regions termed *meristems*, and the zones between meristems and the nearby mature tissues contain cells in intermediate stages of maturation. A comparison of a juvenile and a mature stage illustrates the great precision and specificity with which cell differentiation takes place behind a meristem. Plate 49 shows the central portion of an unusually 'miniaturized' root at an early stage of development and Plate 16a depicts the same cell types, distributed in exactly the same geometrical pattern, but in a mature part of the same root. Six different types of cell have matured in their own characteristic fashions and at their own characteristic rates, all starting from a population of comparatively uniform meristematic cells. In an organization of this sort there is clearly no such thing as a 'typical plant cell', but meristematic cells must at least contain a basic set of components. They alone have not, or have only just, started to diversify by maturation, so it is logical to use them in an introductory survey of 'the cell' (Plates 1–3, Fig. 1), before examining the constituent parts in detail.

Plate 1 is a light micrograph, showing cells in a broad bean root tip that was fixed in glutaraldehyde, dehydrated, embedded in plastic, sectioned at about 1 μm thickness, and the section stained by a combination of procedures chosen to reveal as many as possible of the cell components. Finer detail is visible in Plates 2 and 3, which are electron micrographs of ultra-thin sections of cells in other root tips. The final illustration in the introductory survey (Fig. 1) attempts to overcome the artificial two-dimensional impression created by the micrographs. It is a stylized three-dimensional interpretation of that mythical entity, the 'typical' plant cell. For the sake of clarity it is shown isolated from all the neighbours to which it should be joined. The components drawn within it are somewhat more symmetrical and simplified than they would be in life, also some have of necessity been enlarged in order to make them visible alongside their larger companions.

Plant tissues are composed of the non-living *extracellular* region and the living *protoplasm* of the cells proper. The former consists of *intercellular spaces* and *cell walls*. Each *protoplast* consists of *nucleus* (or sometimes several *nuclei*) and *cytoplasm*, and within these major subdivisions of the cell are the various membranous and non-membranous components with which subsequent descriptions are concerned. The following list gives little more than a list of names and outline descriptions. Functions are not included. The numbers that follow each item indicate which of the first three Plates give the best views of the structure in question; the letters in brackets refer to labels on Fig. 1.

Cell wall: This is a thin structure in meristematic cells, but it can be very massive and elaborate (1, 2).

Plasma membrane: The bounding membrane of the protoplast, normally in close contact with the inner face of the cell wall (2, 3b).

Plasmodesmata: Narrow cytoplasmic channels, bounded by the plasma membrane, and interconnecting adjacent protoplasts through the intervening wall. The singular is *plasmodesma* (2, 3b). (PD).

Vacuole: Compared with the surrounding cytoplasm, these are usually empty looking spaces, spherical when small. They can, however, be very large, exceeding by many times the bulk of the cytoplasm itself (1). (V).

Tonoplast: The membrane that bounds a vacuole. Except for its position in the cell it looks very like the plasma membrane (3a, 3b).

Nuclear envelope: A cisterna (a general term meaning a membrane-bound sac) wrapped around the contents of the nucleus. (N). The space *between* the two membranous faces of the cisterna is the *peri-nuclear space* (1, 2, 3a).

Nuclear envelope pores: Perforations in the nuclear envelope, through which the cytoplasm may be in continuity with the contents of the nucleus (3a). (NP).

Chromatin: The genetic material of the cell, containing the information that is passed from parent cell to daughter cell during the multiplication of cells and reproduction of the organism. It can exist in various forms, and during the division of nuclei it is condensed, so that discrete units, *chromosomes*, can then be recognised (1).

Nucleolus: A mass of fine threads and particles, largely a sequence of identical units of specialized genetic material together with materials—precursors of ribosomes—produced from that genetic information (1, 2). (NU).

Nucleoplasm: Everything enclosed by the nuclear envelope falls in the category of nucleoplasm, just as objects outside it are constituents of the cytoplasm. The word is often, however, used to denote the ground substance in which the chromatin and nucleolus lie (1, 2).

Endoplasmic reticulum: Cisternae that ramify through the cytoplasm, occasionally connected to the outer membrane of the nuclear envelope. The inner face of the bounding membrane is in contact with the contents of the cisterna, and the outer face frequently bears attached ribosomes and polyribosomes (see below). Endoplasmic reticulum is described as 'rough', or granular, (RER) and forms that lack ribosomes as 'smooth', or agranular, (SER). The cisternae may or may not have visible contents, distending the cisternae when present in bulk (2, 3a, 3b).

Ribosomes: Small particles lying free in the cytoplasm or else attached to the endoplasmic reticulum. They may be aggregated in clusters, chains, spirals, or other *polyribosome* configurations (2, 3a).

Dictyosomes: The units of the *Golgi apparatus* of the cell. Each dictyosome consists of a stack of cisternae, and many small vesicles (1–5) (VE).

Mitochondria: Components consisting of a compartment surrounded by a double envelope. The outer membrane of the double envelope is more or less smooth, but the inner is thrown into many folds—*mitochondrial cristae*—that project into the central compartment (2, 3a) (M).

Plastids: This is a group name for a whole family of cell components represented in the root-tip cells of the first three plates by the simplest member, which is called the *proplastid* (P). Proplastids are usually larger than mitochondria, but like them have a double membrane envelope surrounding (in these examples) a fairly dense ground substance. *Starch grains* may be present in them (1, 3a) (ST).

Other members of the plastid family, illustrated in later plates, are: *chloroplasts, etioplasts, amyloplasts,* and *chromoplasts*.

Microbodies: These are bounded by a single membrane, and are distinguished from other vesicles by their size and dense contents (sometimes including a crystal) (2) (MB).

INTRODUCTION

Microtubules: Except during cell division, these very narrow cylinders lie just inside the plasma membrane. The wall of the cylinder is made of protein and is *not* a cell membrane, though it superficially resembles one in its thickness and density (2, 3a) (MT).

Microfilaments: Fine fibrils composed of a material resembling actin, one of the major constituents of muscle. They are not illustrated in Plates 1–3.

Subsequent micrographs and their captions examine these components in more detail, and since many of them are present in meristemic cells in only a very simple form, this will mean exploring specialized cell types as well, in order to seek out and assess significant aspects of the known diversity.

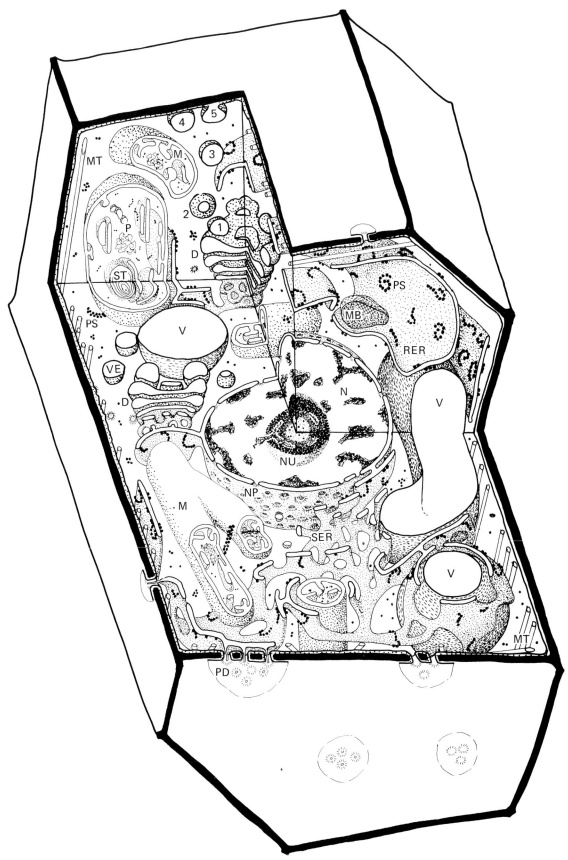

Fig. 1
Diagram of an undifferentiated cell cut open to show the three-dimensional structure of the principal components and their interrelationships. For clarity they are not drawn to scale and some are illustrated by only a few examples (e.g. ribosomes, see Plate 2). They may be identified by the letters, which refer to those in parentheses in the list on pages 5 and 6.

Plate 1 (See notes at foot of page)

The Plant Cell (1). Light Microscopy

These large cells in the meristematic region of a broad bean (*Vicia faba*) root tip are viewed by phase contrast microscopy. The section was reacted with acriflavine, following oxidation in periodic acid, to stain carbohydrates yellow (e.g. in cell walls and starch grains), and subsequent immersion in another yellowish reagent—iodine in potassium iodide—gave a generally-stained preparation, best examined using blue light. The magnification is × 4200, i.e. 4.2 mm represents 1 μm, and since the thickness of the section was about 1 μm, we are in effect looking through a slice 4.2 mm in thickness, rather than at an infinitely thin 2-dimensional picture.

Each cell is outlined by its wall (CW). There are no intercellular spaces in this particular group of cells. Within the walls the major visible compartments of the cells are the numerous empty-looking vacuoles (V), the cytoplasm around the vacuoles, and the nuclei, which are numbered N1 to N4.

The nuclei are separated from the cytoplasm by the nuclear envelope (NE), best seen in nucleus (N-1), which has been caught at an early stage of division. In the other nuclei the speckles represent stained chromatin (CH); prior to division this condenses to form discrete chromosomes (CHR in N-1), leaving the nuclear envelope relatively isolated and conspicuous. After division the chromosomes uncoil again to regenerate the chromatin condition, and a stage of this process is seen in nucleus N-2. The large dense bodies in the nuclei are nucleoli (NL). The nucleolus in N-1 is lobed and irregular, but in the non-dividing nuclei its circular outline is indicative of a more-or-less spherical shape. Pale nucleolar 'vacuoles' occur in most nucleoli. No nucleolus is seen in nucleus N-3, but this does not mean that none is present. Sections are statistical samples of cells and tissues, and it is not to be expected that any one view of a cell will contain all of the components. Consider the dimensions of nucleus N-4, and assume that it and its nucleolus are spheres, both sectioned across their diameters, which at 4.2 mm representing 1 μm, are about 10 μm and 4 μm respectively. It would take 10-11 consecutive 1 μm sections to pass from one face of the nucleus to the other, and only 4-5 of these would include portions of nucleolus.

The most clearly resolved cytoplasmic components are proplastids (PP), but these can only be identified with certainty if starch grains (e.g. arrow, top right) can be detected inside them. It would appear from their varied profiles in the section (some elongated, some less so, some mere dots) that there is a population of randomly oriented, more-or-less sausage shaped proplastids in the cells. Only rarely does the full length of a 'sausage' lie within the thickness of the section. Other cytoplasmic structures can be discerned but they cannot be identified with certainty. The less densely stained particles doubtless include mitochondria (M?), and the very faint convoluted shadows (e.g. connecting the small arrows above N-1) are probably cisternae of endoplasmic reticulum, but this is a statement which can only be made with the benefit of hindsight by an observer who has studied these and similar cells with the electron microscope.

Notes: (1) Page numbers at the foot of each plate caption correspond to those in *Ultrastructure and the Biology of Plant Cells* (see inside front cover).

(2) Numbers in the margins of the plate captions refer to text page numbers in *Ultrastructure and the Biology of Plant Cells*.

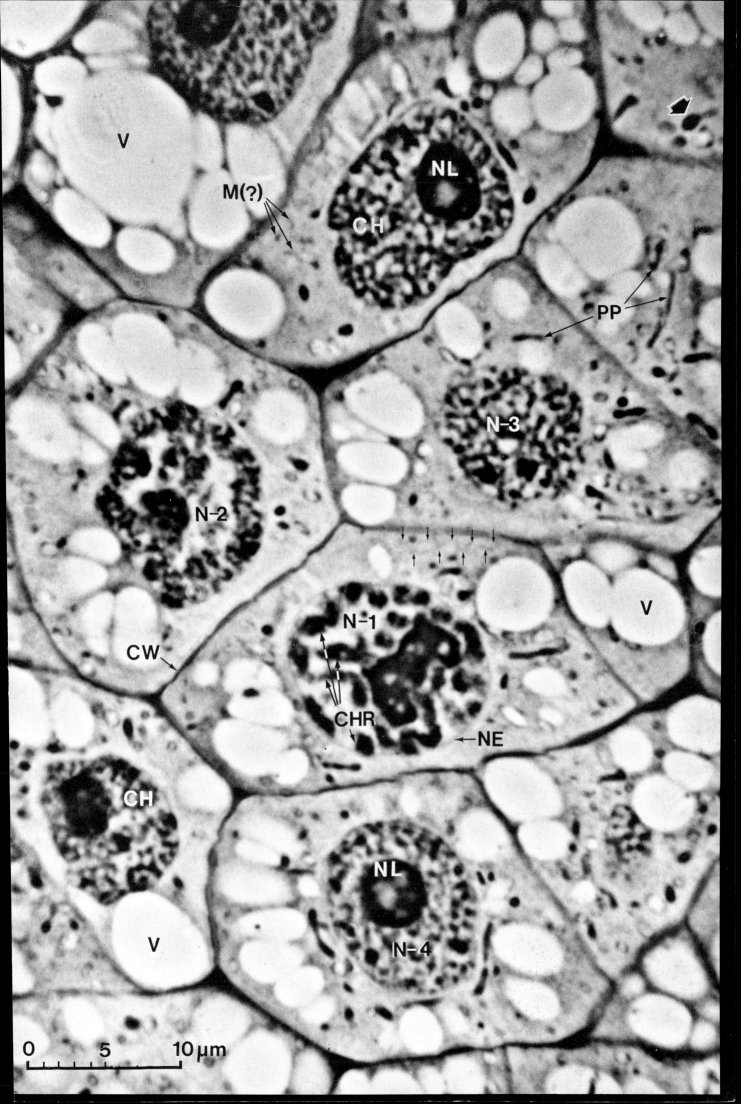

Plate 2

The Plant Cell (2)

This section sliced through the mid-region of a cell in a root tip of cress (*Lepidium sativum*), and is viewed here by electron microscopy at × 20 000. The section was about 75 nm thick, so at this magnification the slice is about 1.5 mm thick—relatively thin compared with Plate 1.

The resolution of the electron microscope brings to light many features not seen in Plate 1. The plasma membrane (PM) is clearly distinguishable from the wall external to it (CW). By and large the plasma membrane lies at right angles to the plane of the section, so that we are looking at it edge-on. It therefore appears as a dark line in the section. The plasma membranes of adjacent cells pass through the intervening cell wall at plasmodesmata (PD). If we could see these extensions of the plasma membrane end-on, as in a section cut in the plane of the wall, they would appear as round holes. Again remembering that the section is of finite thickness, we can estimate roughly how many plasmodesmata pierce the wall. There are, for instance, 11 in the 250 mm of the left hand upright wall, and since at × 20 000, 20 mm represents 1 μm, this means 11 in an area of wall equal to 12.5 μm multiplied by the section thickness, i.e. about 75 nm. From this we can calculate that there are up to 1200 plasmodesmata per 100 μm² of this type of wall.

The vacuoles (V) are sparse and small in this view (cf. Plate 1). They lie in a densely particulate cytoplasm, each tiny particle being a ribosome. Ribosomes either lie free or else are attached to the membranes of the rough endoplasmic reticulum cisternae, giving them a characteristic beaded appearance (ER). In these cells most of the cisternae are narrow, and it should be realized that only those which lie at right angles to the plane of the section or nearly so are clearly visible.

Other components of the cytoplasm are present: mitochondria (M), proplastids (PP), dictyosomes (D), and a microbody (MB). Note the difference in the electron-density of mitochondria and proplastids. A small region (outlined in black at top left) is shown at magnification × 40 000 in the insert (lower left) in order to make the cross sections of microtubules present there more obvious (arrows).

The inner and outer membranes of the nuclear envelope (NE) are resolved. The outer, like the endoplasmic reticulum, bears ribosomes. The nuclear contents, perhaps most conspicuously different from the cytoplasm in the absence of membranes, consist of nucleolus (NL) and chromatin (CH) suspended in the general ground substance, or nucleoplasm. The nucleolus is an unusually large example, with fewer and smaller 'vacuoles' than those in Plate 1. It exhibits differentiation into regions that are predominantly granular (solid stars) and others in which substructure cannot be seen at this magnification, being composed of closely packed fine fibrils (open stars). The open arrows point to portions of strands of chromatin that ramify through a network of channels in the body of the nucleolus.

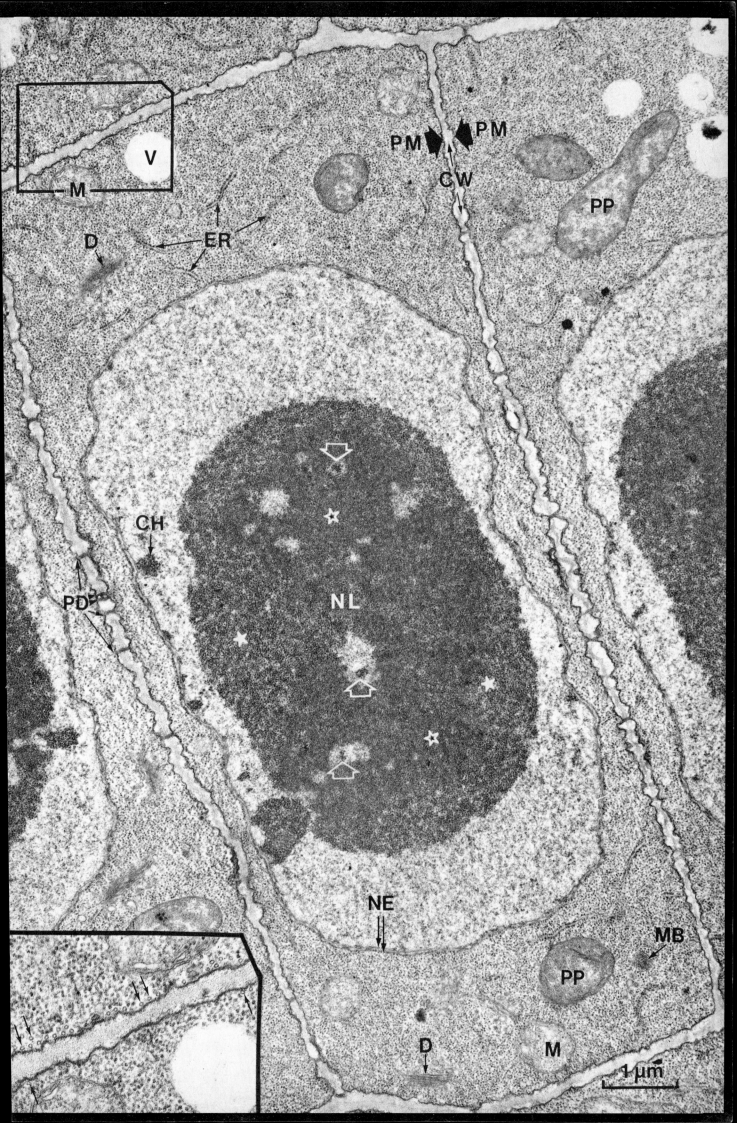

Plate 3

The Plant Cell (3)

Plate 3a The same material as in Plate 2 is seen here at a higher magnification ($\times$ 45 000). Parts of two cells are shown, separated by a wall (CW), lined by the two plasma membranes (PM). Some microtubules (arrows) lie just internal to the plasma membrane, and two (circled) are unusually deep in the cytoplasm. The cisternae of endoplasmic reticulum (ER) are all 'rough', with ribosomes studding their membranes and also lying free in vast numbers. One cisterna (ER*, lower right) is distended by the presence of accumulated contents. The left hand cell is not as free of endoplasmic reticulum as might appear at first glance. It is merely that many of its cisternae happen to lie in or close to the plane of the section. The arrowheads indicate the limits of one, barely discernible, obliquely-sectioned cisterna.

A small portion of a nucleus (N) is included at the right hand side, and the two membranes and pores (open arrows) of the nuclear envelope (NE) are visible. The successive cisternae of the dictyosome (D) lie at right angles to the plane of the section, and there are many associated vesicles nearby in the cytoplasm, or else attached to the cisternae. The double envelope of both proplastids (PP) and mitochondria (M) can be seen (squares). The tonoplasts of the two vacuoles that (only just) enter the lower edge of the picture again illustrate the difference between the crisp profile of a membrane lying 'edge on' (T) and an indistinct, obliquely sectioned membrane (T*).

Plate 3b This high magnification picture ($\times$ 100 000) illustrates one aspect of the substructure of cell membranes. The material (a young cortical cell of the root tip of *Azolla*) is especially instructive for two reasons. One is that the cells contain some substance which spreads (possibly during fixation) to the surfaces of all membranes and highlights their dark-light-dark appearance when seen edge-on in an ultra-thin section. The other is that the cells accumulate much lipid, and in older examples the vacuole (V) contains massive deposits. A small droplet of lipid in the form of a myelin figure (MF) is half way between cytoplasm and vacuole. It shows concentric layering, formed by back-to-back alignment of bimolecular layers. The cell membranes themselves show somewhat similar layering, but because they are either isolated, or else not stacked as closely as the lipid, the dark-light-dark 'tramlines' are seen individually.

The plasma membrane (PM) lines the cell wall (CW) and pierces it at plasmodesmata (PD). Other membranes included are the tonoplast (T), and cisternae of endoplasmic reticulum (ER) and a dictyosome (D).

The inserts show enlarged details from the same micrograph ($\times$ 250 000). From left to right: bimolecular layers of lipid in the myelin figure; plasma membrane; tonoplast; endoplasmic reticulum; and dictyosome cisterna. In each case the circles enclose one short portion of triple-layered membrane.

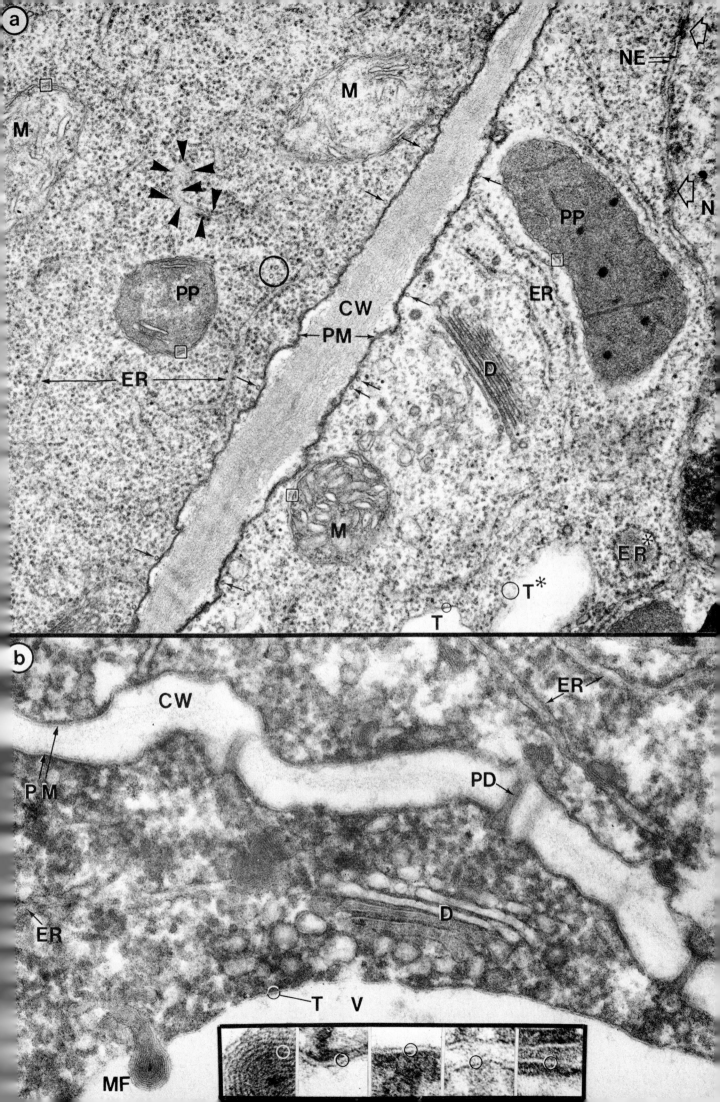

Plate 4

Plasma Membrane, Microfibrils in the Cell Wall

Plate 4a This face view of part of the plasma membrane (upper left) and cell wall (lower right) of a cell in an *Asparagus* leaf was obtained by the freeze-etching technique. Layers of microfibrils can be seen in the wall, and the scattered dark holes may be plasmodesmata (PD). Many particles about 10 nm in diameter lie in the plasma membrane, sometimes in rows (e.g. arrow). The particulate texture of some areas (asterisks) is different from the surrounding membrane. Semi-crystalline patterns can occur (not seen here). At one place the whole thickness of the plasma membrane has ripped away to expose another membrane surface within the cytoplasm (star). The particles in the expanse of membrane are probably proteinaceous, while the smooth areas between the particles represent an internal surface composed of lipid molecules, exposed when the membrane fractured along its mid line.

Magnification × 95 000. (We thank Drs. H. Richter and U. Sleytr and Verlag G. Fromme & Co., Wien, for permission to produce this micrograph from *Mikroskopie*, **26**, 329-346, 1970.)

Plate 4b and c The cellulose microfibrils illustrated here by means of shadow casting are in pieces of primary cell wall from which matrix materials have been extracted (cf. Plate 4a). Plate 4b is a wall viewed from the plasma membrane side. The most recently deposited microfibrils have not been pulled far out of their original orientation (which corresponds to the axis at right angles to the long axis of the cell) but the older, deeper microfibrils are more randomly oriented. The scattered sites of individual plasmodesmata are neatly circumscribed by curved microfibrils (PD) (elongating pith cell of *Ricinus*, × 22 000). Plate 4c shows (i) randomly arranged microfibrils (top centre), (ii) specific delimitation of pit fields where plasmodesmata are clustered (asterisks, see also Plate 14e), and (iii) parallel orientations (running across the lower part of the micrograph) at a strengthening rib, all in an oat coleoptile cell (× 15 000). (We thank Dr. K. Muhlethaler and Buchler und Co. AG. for permission to reproduce these plates from *Ber. Schweiz. Botan. Ges.* **60**, 614, 1950).

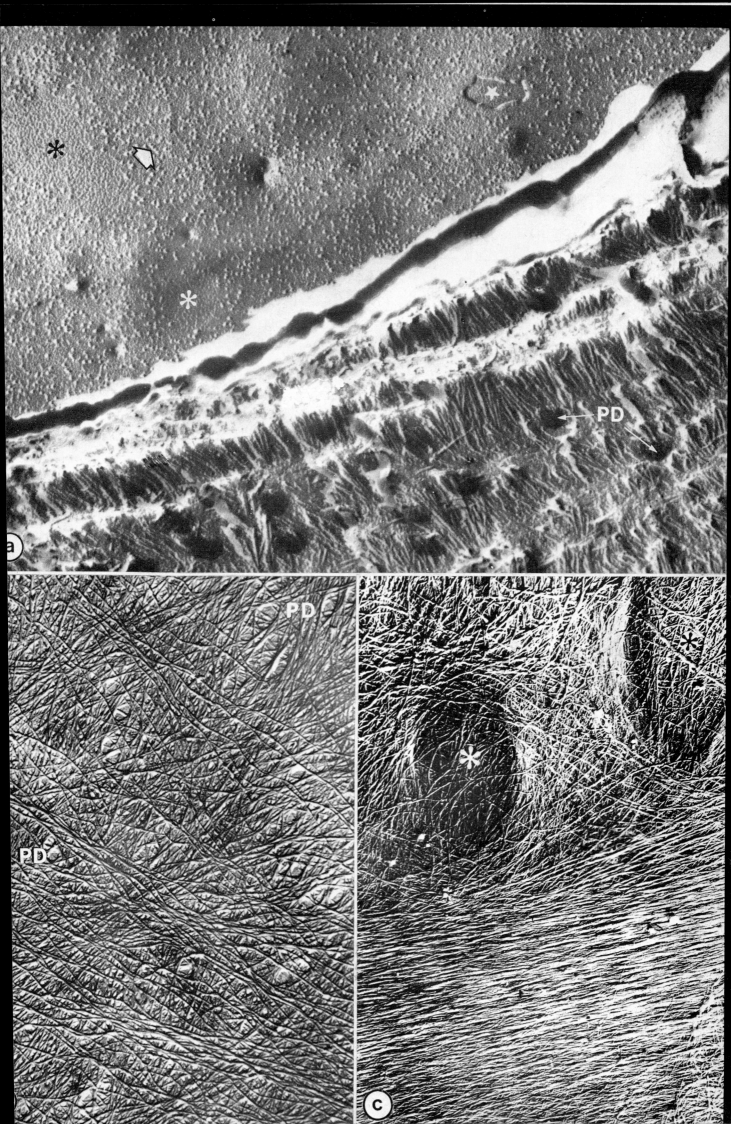

Plate 5

Xylem (1): Developing Xylem Elements

Xylem is the tissue which carries the transpiration stream from root to shoot. It consists predominantly of dead xylem elements that are aligned in rows to form vessels or tracheids, together with the living xylem parenchyma cells responsible for transferring solutes into and out of the non-living conduits. Developing xylem elements (not yet dead and empty) are shown in Plate 5, and mature systems are illustrated in Plate 6.

Plate 5a This micrograph shows a longitudinal section of xylem elements in the vascular tissue of an *Azolla* root (see Plate 49 for transverse section). PX is a protoxylem element, recognizable as such because the lignified rings have been pulled apart (arrows) by elongation of the surrounding tissues after the protoxylem was itself mature. The black areas (asterisks) represent parts of cells that bulge into the protoxylem and so have been included in the plane of the section. MX is a metaxylem element, developing much later than the protoxylem. The many wall thickenings are seen (e.g. arrows), as is an end wall (W) separating one xylem element from the next in the file. Nucleus (N) and cytoplasm are still present. × 4300.

Plate 5b Part of the wall between two developing xylem elements is illustrated here in a section of leaf vein. The uppermost cell still has its cytoplasm: three dictyosomes (D) and their associated vesicles are prominent. The tonoplast (T) is intact. Scrutiny of the same field of view at higher magnification showed the position of the microtubules which overlie developing xylem thickenings. Since the microtubules are too small to be seen here, their positions are marked by arrows, 34 in all. Their association with the lignified bands is very clear.

The lower cell is more advanced in its development. Its tonoplast has ruptured and only a few recognizable cytoplasmic components remain. Although its cytoplasm has been largely digested, breakdown of the primary wall has not yet commenced. All but the cellulose microfibrils will become digested from zones between the thickenings, to give a mature condition as in Plate 6c. Magnification × 16 500. *Lupinus* leaf.

Plate 5c and d Developing xylem in *Phaseolus* (French bean) at × 1800, viewed by light microscopy. Plate 5c shows the onset of lignification in the thickenings (pale areas, L, contrasting with the darker stained non-impregnated hemicellulosic parts (H) of the young thickenings). The primary wall (PW) is still present, with some cytoplasm. Plate 5d shows a later stage, with no cytoplasm, loss of the primary wall (PW) except where protected by the lignin, and uniformly lignified thickenings. Note the back to back arrangement of the thickenings in neighbouring cells.

Plate 5e Developing xylem of *Galium* (goosegrass), viewed in transverse section at × 13 500, shows many of the features seen in Plate 5a and the upper half of Plate 5b, including microtubules (arrows) at the developing thickenings. In this plane of section the microtubules are seen in side view rather than transversely cut. There are several dictyosomes (D) and numerous vesicles (thought to originate from the dictyosomes) in the cytoplasm.

Plate 5f, g and h Three stages in the dissolution of the end wall (W) of vessel elements in *Phaseolus* are seen here by light microscopy at × 900. The mature vessel (derived from vessel elements) is thus an open pipe, through which the transpiration stream flows from root to shoot.

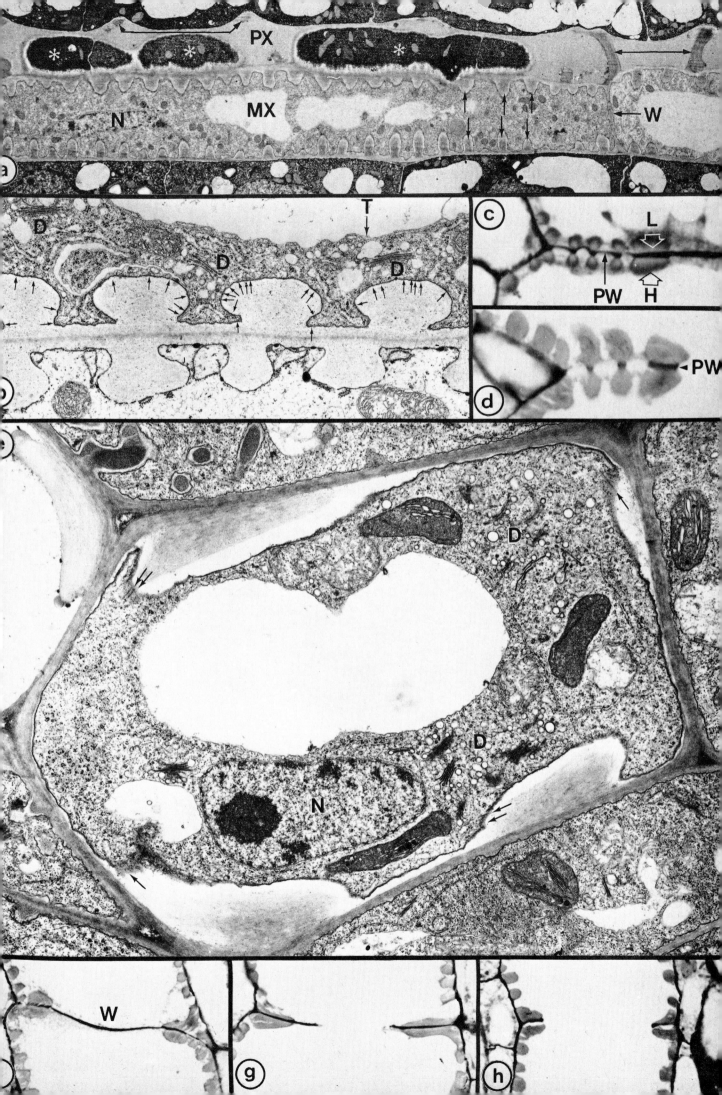

Plate 6

Xylem (2): Mature Xylem and Xylem Parenchyma

Plate 6a Part of a long xylem element, isolated from a lettuce leaf vein by digestion with snail gut enzymes, is viewed here by scanning electron microscopy at × 3250. Except in a few places the remains of the primary wall (PW) have been digested away, and only the reticulate lignified thickenings (L) remain. Snail digestive juice contains enzymes that destroy cellulose and other wall carbohydrates.

Plate 6b In this scanning electron micrograph (× 3400) xylem elements in a pea leaf vein that was partially isolated after treatment with the enzyme pectinase are viewed (as in 6a) from the outside, though one can see into the lumen at the broken end of the topmost element (white arrow), where both outside and inside faces of the lignified thickenings are visible. The continous wall between the thickenings has suffered tearing (asterisks) during specimen preparation. It is probably composed largely of cellulosic remnants of the primary wall, and is retained here more than in 6a because of the absence of cellulose-digesting enzymes during preparation.

Plate 6c In this ultra-thin section (× 7200) of part of two xylem elements in a pea leaf vein, living xylem parenchyma cells just enter the picture at top and bottom. The lignified thickenings and the wisps of cellulose microfibrils (arrows) that interconnect them are the only components to survive the auto-digestion processes of xylem maturation. These wisps presumably constitute the expanses of continuous wall seen at much lower resolution in Plate 6b. The thickenings of the upper element are much more widely spaced than those in the lower, probably indicating that it matured earlier, and became passively stretched, as in the protoxylem seen in Plate 5a.

Plate 6d This transverse section of primary xylem tissue in the stem of a fumitory (*Fumaria*) seedling includes the following features of xylem elements (X): lignified thickenings, cellulosic remnants of primary wall in exposed localities (arrowed circles) and more complete survival in areas *underlying* lignified thickenings (arrows). An important feature not previously illustrated is that living xylem parenchyma cells lie around and between the xylem elements. They are responsible for loading the xylem sap with solutes such as mineral nutrients, amino acids, and some hormones, and also for absorbing solutes from the xylem sap. Note the many cisternae of endoplasmic reticulum in the xylem parenchyma cells. Such cells can be shown to absorb amino acids from the xylem sap and to manufacture protein. Magnification × 2700.

The inserts show scanning electron micrographs: of primary xylem elements comparable to those in the section, with helical lignification (insert at lower left, × 2200); and of secondary xylem, where the lignification is much more extensive, and covers the whole wall except at lens-shaped pits (insert at upper right, × 2200). (Both inserts are of castor bean (*Ricinus*) stem, and were kindly provided by Drs. J. Sprent and J. Milburn).

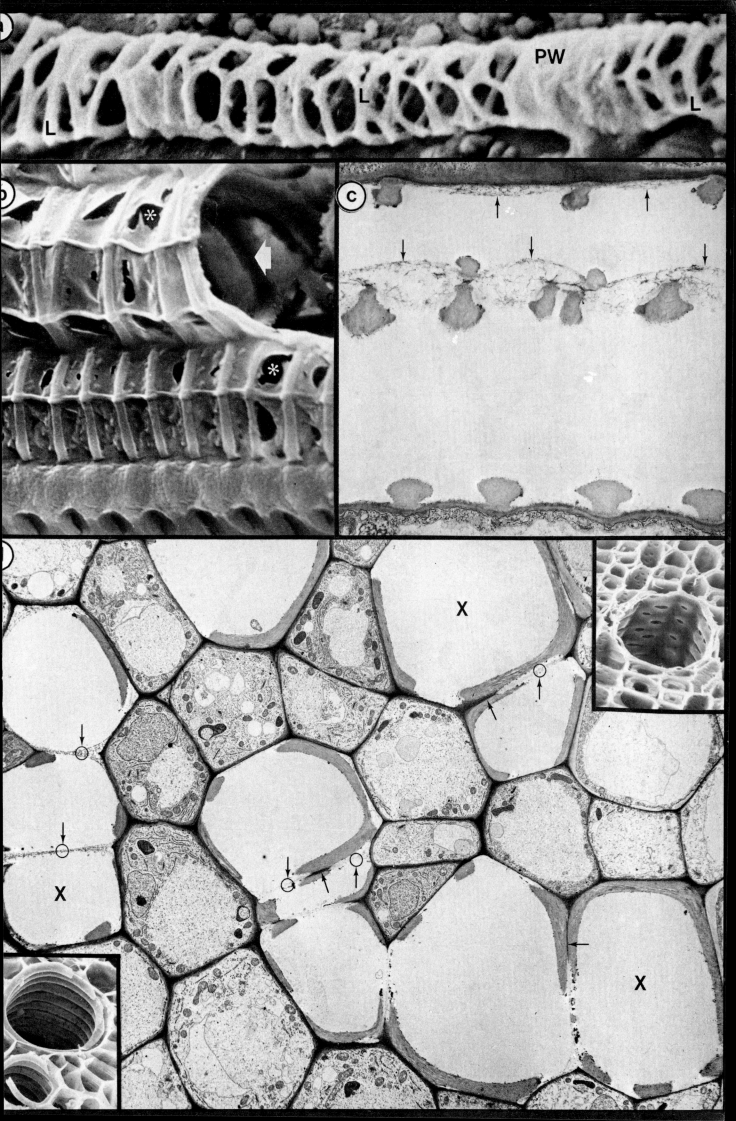

Plate 7

Phloem (1): Sieve Element and Companion Cell

Phloem is the tissue responsible for long distance transport of metabolites, principally sugars, from sites of production to sites of consumption in plants. This micrograph illustrates in cross section the two principal types of cell concerned with the transport process. They are the sieve elements (lower cell) which are *apparently* empty conducting cells, and the parenchyma cells, some of which, the companion cells (upper cell) are derived from the same parental cells as the sieve elements. In flowering plants, mature sieve elements are interconnected in series to form sieve tubes (Plate 8).

Sieve elements mature following the breakdown of their tonoplast and much of their cytoplasm (Plate 24). The lumen of the sieve tube then contains remnant mitochondria (M) and plastids (P) and a system of protein fibres ('P-protein', arrows). In contrast to xylem cells (Plates 5 and 6) the plasma membrane of the sieve tube (PM) remains intact, in fact transport will only occur if this membrane survives. Its maintenance is one of the suggested functions of the companion cells, whose own plasma membrane is continuous with that of the sieve tubes *via* numerous plasmodesmata (PD). Plasmodesmata between sieve elements and companion cells are typically compound (see also Plate 14h), and incorporate linings of electron-transparent callose (C). The plasmodesmata may provide a route whereby the companion cells can function in the loading and unloading of sugars to and from the sieve elements at different points around the plant. Companion cells may also provide energy for the transport process itself (in addition to maintenance of the sieve tube plasma membrane). They are richly endowed with mitochondria.

The relative size of the two cell types differs according to the nature of the transport system. Where loading predominates over longitudinal movement, as in the minor veins of leaves, which drain photosynthates from the mesophyll tissue, the companion cells are of much greater diameter than the sieve elements. Conversely, in stems, where longitudinal movement is the major process, it is the sieve elements which are by far the larger. The example shown here is intermediate in character. It depicts part of the phloem in a vascular strand in the ovary of the grape hyacinth, *Muscari*, × 3000.

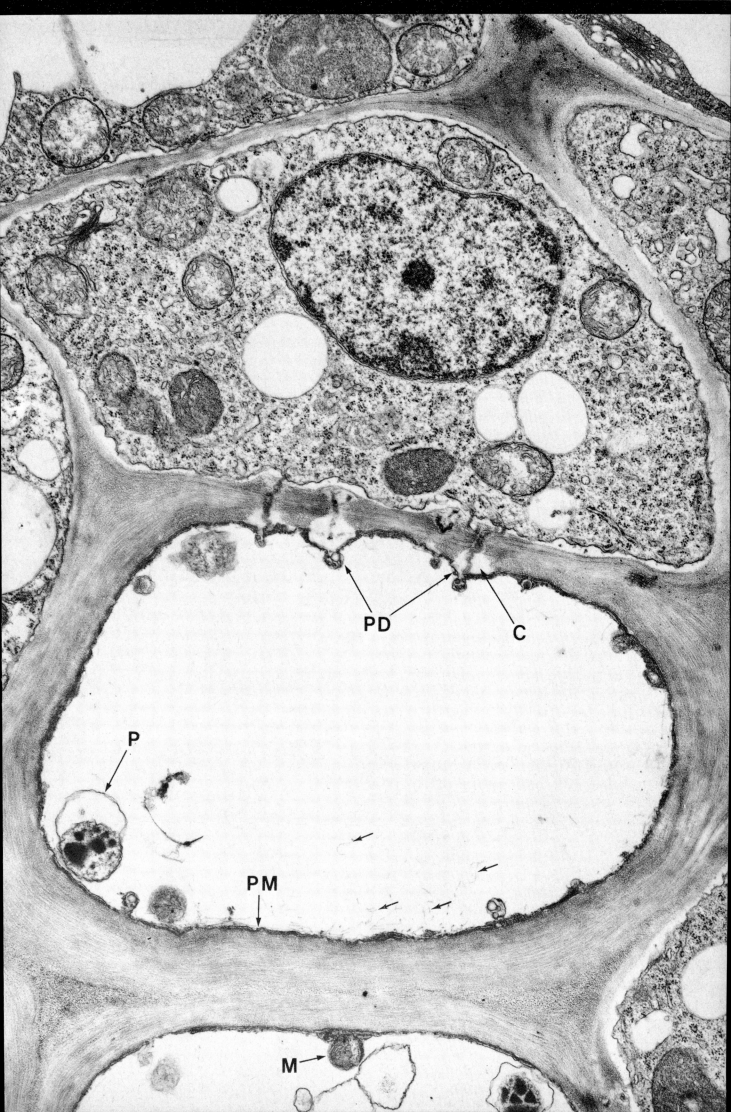

Plate 8

Phloem (2): Sieve Plates and Sieve Pores

Plate 8a In longitudinal section sieve tubes (ST) are seen to be interrupted at intervals by perforated cross walls, the sieve plates (SP), which were originally the walls between successive cells in the file of differentiating sieve elements. The pores develop from plasmodesmata in the primary cross walls. In this particular tissue some of the accompanying companion cells have become specialized by the formation of wall ingrowths (see also Plate 13). Note the abundance of plasmodesmata between companion cells and sieve tubes (arrowheads). Such plasmodesmata may be the paths through which the sieve tubes are loaded or unloaded. White lupin petiole, × 3500.

Plate 8b, c and d It is possible that none of the pictures presented here is a faithful representation of the sieve plate of a functional sieve tube. Sieve plates and their pores present a number of different appearances in thin sections due to changes that take place after material has been sampled from the intact plant and during the subsequent fixation period. These changes are brought about by injury-response systems, probably evolved as being advantageous in the prevention of leakage of solutes from the sieve tubes after wind or insect damage to the plant. In the condition thought to be closest to the normal state (8b) only a very small amount of callose (C, the electron-lucent ring around each pore) is present and the pores contain relatively few P-protein fibrils (P). Since sieve tubes are under considerable hydrostatic pressure, due to the osmotic effect of their sugary content, damage to the system results in a sudden release of pressure and a surge of liquid down the sieve tube. This carries masses of the P-protein fibrils (normally dispersed throughout the sieve tube) on to the sieve plates and into the pores (8c). A temporary blockage of the pores is thus effected while another reaction takes place, the formation of massive callose plugs which seal the pores (8d). Callose formation is extremely rapid and the response system is sensitive to chemical as well as to mechanical damage to the phloem. Some insects, notably aphids, do manage to circumvent these response systems, probably by creating only a small pressure drop in the parts of the sieve tubes tapped by their stylets, so they can successfully rob the plant of its food supplies. A specialized form of endoplasmic reticulum found in sieve elements is marked ER in (b), (see also Plate 24c). Same tissue as in 8a. 8b, × 27000; 8c, × 16000; 8d, × 14000.

Plate 8e Face view of part of a sieve plate in a similar condition to that in 8a, that is with a copious formation of electron-transparent callose pinching off the pores, which are in addition occluded by P-protein fibrils. The weft of microfibrils that delimit the pores in the sieve plate can be seen. *Coleus blumei* stem, × 19000.

The structure of the sieve plate, and the state of its pores, are key factors in considerations of the mechanism of translocation in the phloem. At the plates, the available pathway for movement is not the lumen of the sieve tube, but the sum of the lumina of a set of very much smaller pores, each one occupied by fibrillar protein in an amount that is difficult to gauge. Longitudinal flow of sugar (etc.) along sieve tubes must be constrained more at the sieve plates than elsewhere. Are they, however, more than mere obstacles to translocation? Provision of a plugging mechanism is one likely *raison d'etre*, but some authors postulate a deeper significance, as sites at which a motive force for phloem translocation might be generated or applied to the sieve tube contents.

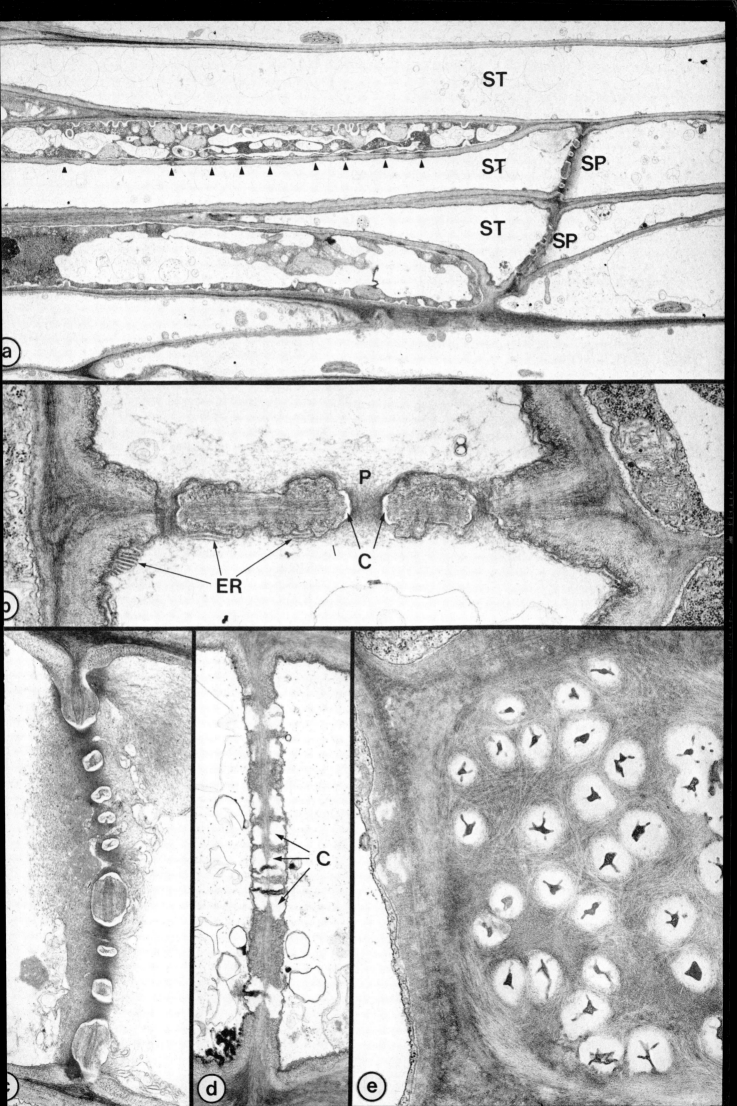

Plate 9

Wax and Cuticle

The outer surface of the plant is usually covered by a layer of protective material that is synthesized by the outermost cells. This plate illustrates some of the range of form exhibited by waxes and cuticles, and shows that they are not confined to the exterior of the plant, but can also develop on the surface of internal cells.

Plate 9a and b Wax formations on the lower surface of a banana leaf (scanning electron microscopy; 9b, × 320; 9a shows the central region of 9b magnified to × 1500). Curled threads of wax completely clothe the epidermal cells, except for the outer cuticularized caps of the stomatal guard cells around the slit-shaped stomatal apertures. The extruded wax threads are 25-50 µm long. (We thank Dr. P. J. Holloway and Mr. E. A. Baker of the Long Ashton Research Station, University of Bristol, for these micrographs).

Plate 9c and d Ridged cuticles on epidermal cells, viewed by ultra-thin sectioning (9c) and scanning electron microscopy (9d). The ultra-thin section in 9c shows the cuticle layer (C) external to the microfibrillar wall (W). The 3-dimensional shape of the ridges is better displayed in 9d, at upper right and upper left. A smooth cuticularized cover of a pair of guard cells is the major feature of 9d, with the central, slit-shaped stomatal aperture. The ability of individual cell types to regulate the form of their adcrusting layers is highlighted by comparing the specific patterns of wax distribution and cuticle morphology shown by guard cells and epidermal cells (9a-d). 9c, daffodil flower corona, × 22 000; 9d, *Auricula* sepal, × 3000.

Plate 9e and f This section (9e, part of a hair from the dead-nettle, *Lamium*, × 15 000) and scanning electron micrograph (9f, a hook-shaped clinging hair from the goosegrass, *Galium aparine*, × 1000) show another common type of cuticle morphology, in which the outer surface is thrown into warty projections. Note also the abundance of longitudinally oriented endoplasmic reticulum cisternae in the section.

Plate 9g The wall seen here is lining one angle of an intercellular space (IS) in the floral nectary of the broad bean (*Vicia faba*). It is heavily cuticularized. Since these cells secrete nectar, it is interesting to note the presence of very numerous branched channels leading from the microfibrillar part of the wall (W) through the cuticle layer (C) towards the intercellular space. In this nectary the secreted sugar solution emerges from intercellular spaces to the exterior *via* modified stomatal apertures. Magnification × 16 500. (Micrograph provided by R. F. Brightwell.)

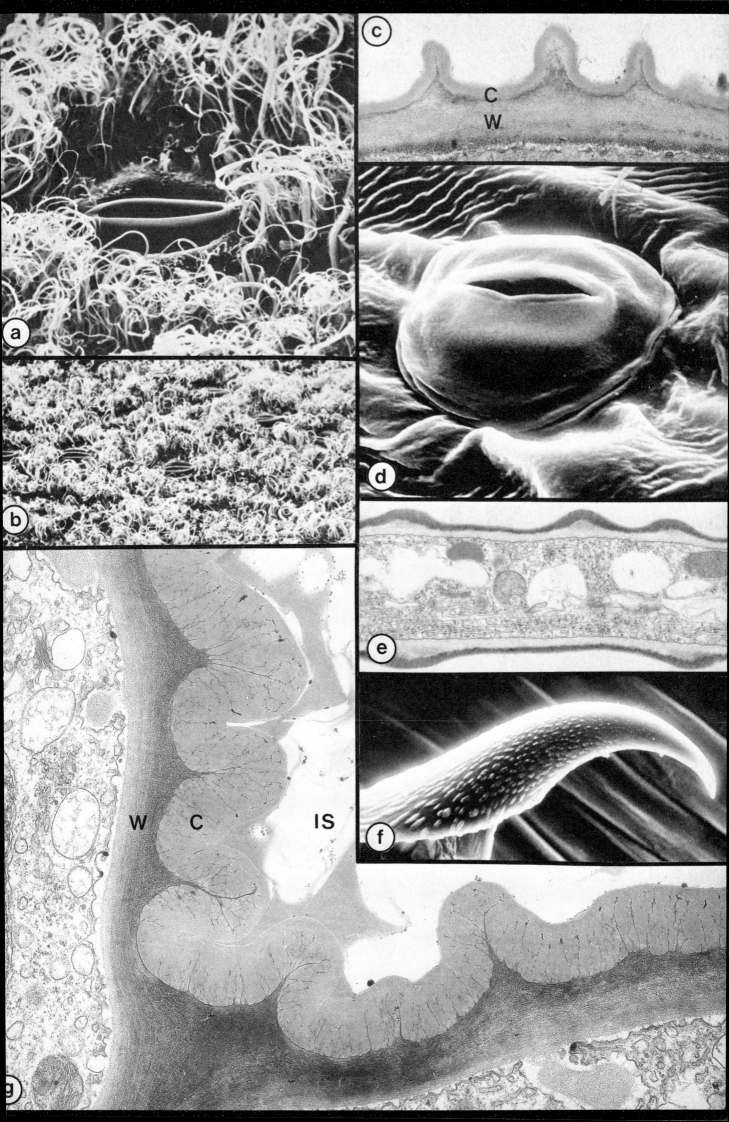

Plate 10

A Capitate Gland

The cell wall covering gland cells and the stalks of glands often shows specialized features.

Plate 10a This gland was found on a young leaf of a dead-nettle (*Lamium*). The secretory cells surmount a stalk cell (S), which itself sits on an extension from a modified epidermal cell. Whilst variation from plant to plant in the chemical nature of substances that are secreted is reflected in corresponding variation in the degree to which *cellular* components are developed, the overall morphology seen here seems to be typical of many glands. They occur on almost all conceivable surfaces of above ground parts of plants. Another example is shown in Plate 23.

Raw materials for the secretion enter the glands *via* the numerous plasmodesmata on the walls of the subtending cell and the stalk cell (arrows). The product passes through the inner, microfibrillar zone of the cell wall of the secretory cells (W), and it is common to find that the cuticle (C) is detached from this layer (asterisks). Presumably the product accumulates in the sub-cuticular space prior to its final exit, which may be through specifically located pores (not seen here) or a more general but slower percolation. × 4800.

Plate 10b The cuticle over the gland is continuous with that (CE in 10a) of the epidermal cells. A specially impregnated zone in the side walls of the stalk cell (stars in 10a) prevents leakage of the secretory product through the microfibrillar layer of wall back down to the subtending epidermis. This wall region is shown at higher magnification in 10b. Whether the impregnation is of cutin or suberin is open to question. Two groups of microtubules (arrows) lie at the extremities of the impregnated zone—an observation which raises the question of whether they might, earlier in the development of the gland, have had anything to do with specifying the distribution of whatever cytoplasmic components were responsible for synthesizing and/or depositing the impregnation. At any event, it seems clear that in this region there is no *non*-impregnated microfibrillar wall outside the plasma membrane (PM), through which secreted products might leak, or through which water might escape and evaporate from the leaf to the exterior. × 30 000.

Plate 10c This micrograph shows details of part of the surface of one of the secretory cells, including the cuticle and the underlying microfibrillar layer, with the space between the two. The plasma membrane (PM) undulates, and, as is common in secretory and absorptive cells, cisternae of endoplasmic reticulum (ER) approach very close to it (arrows). This close juxtaposition may well be related to the transport of solutes between internal compartments of the cell and the extra-cytoplasmic spaces. In other gland cells, marked changes in the disposition of endoplasmic reticulum cisternae occur when secretion is initiated or ceases. Dictyosomes (D) are also present. × 70 000.

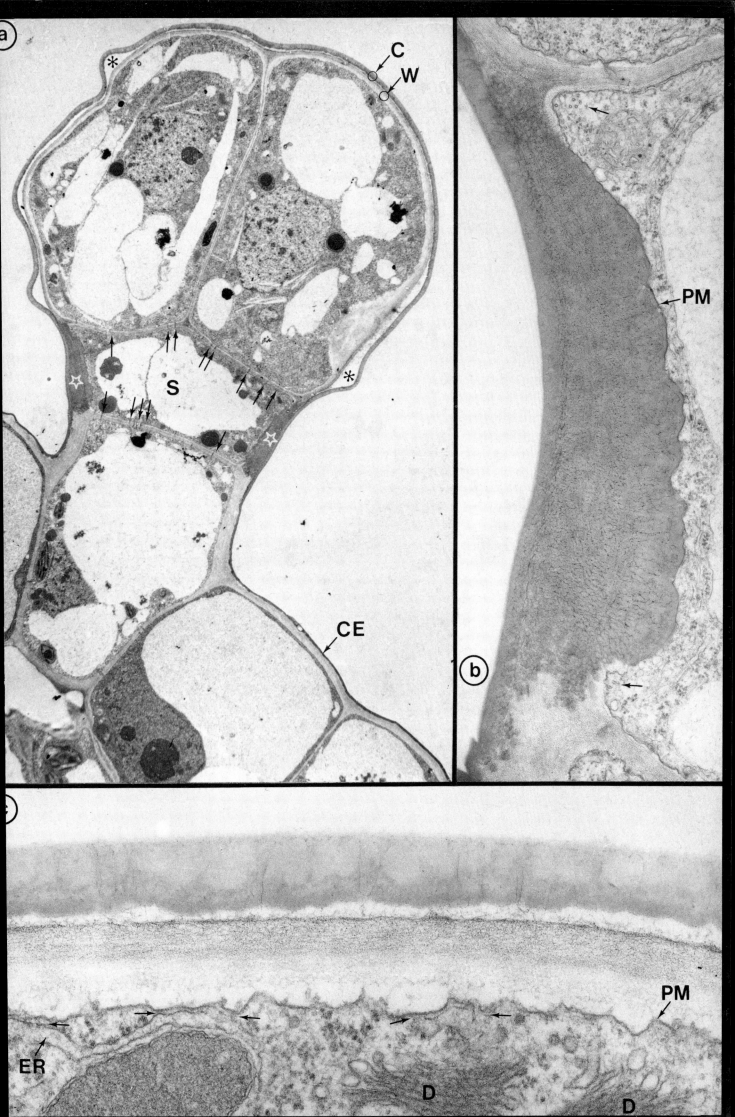

Plate 11

Pollen Grains (1): Developmental Stages

Plate 11a In a very early stage of pollen development, the pollen mother cells of the sporogenous tissue of the anther become isolated by deposits of callose (C) so massive that the primary wall is obliterated, apart from the middle lamella (circles). As in sieve plates and plasmodesmata (Plates 7, 8, 14h) the callose appears virtually structureless and electron-transparent. Initially the pollen mother cells are interconnected by cytoplasmic bridges (PD*), facilitating rapid distribution of nutrients and synchronizing the development of the cells. Remnants of true plasmodesmata (PD) are also present. Both types of intercellular connection are eventually sealed by callose deposition. × 22 000.

Plate 11b and c As in 11a, these micrographs show developing pollen of the cultivated oat (*Avena sativa*), but at a more advanced stage of development in which the callose walls have been digested away and the mature walls laid down. 11b is an enlarged (× 16 500) detail from 11c (× 2800). In 11c, maturing pollen grains are seen in the anther loculus (L), the wall of which consists of three layers of cells (stars) (the innermost layer partially crushed). The loculus is lined by a layer of nutritive cells, the tapetum (T) (considered in detail in Plate 22).

The tapetal cells have almost lost their primary walls, leaving their plasma membrane exposed (PM in 11b). Their nuclei are typical of many cells of very specialized function in that the bulk of the chromatin is extremely condensed, appearing electron dense. It is thought that the remaining dispersed chromatin controls the specialized functions of the cells, which in this case primarily concern the production of nutrients for pollen grain development, and the production of a carpet of orbicules. These irregularly spherical bodies (arrows, 11b) are seen lining the anther loculus on the tapetal cells. They become coated with the very resistant compound sporopollenin. Superficially they resemble units of the pollen grain wall, but there is evidence that the latter is manufactured *in situ*, and *not* by transfer of orbicules from the tapetum as 'building blocks'.

The pollen grain wall is made up of a thin layer called the intine (I), of microfibrils and matrix, together with the exine (E), composed largely of sporopollenin. The exine is subdivided into an inner platform, the nexine (N) from which rods, or bacula (B), extend like columns. A roof, or tectum (T) surmounts the bacula. Oat pollen grains, like various others, have a single, circular pore (see 11b), where the exine bulges in a rim around a separate lid, or operculum (O), of sporopollenin sitting on a thickened pad of intine. This pad in particular, and the intine in general, are major sites of the hayfever antigens that diffuse rapidly out of moistened, mature pollen grains. The pore is also the site of pollen tube emergence during germination of the grain, and it may also function in the water economy of the grain—opening in moist conditions and remaining tightly shut in dry conditions.

At the stage illustrated the pollen grains are vacuolated and have not yet accumulated food reserves. They have not long been released from the tetrads formed at meiosis. Their irregular shape may be an artefact. The pore arises on that face of each grain that lies outermost in the tetrad—a morphogenetic feature that recurs in a somewhat different form in Plate 12a.

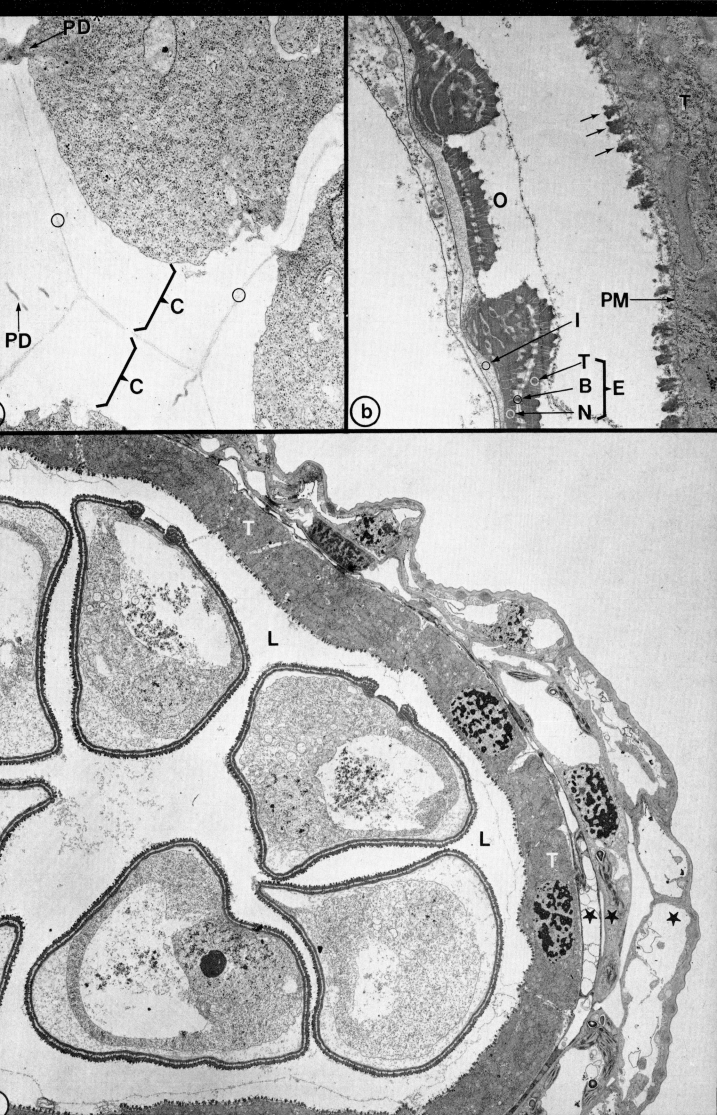

Plate 12

Pollen Grains (2): The Mature Wall

The scanning electron microscope is an ideal instrument for examining the suface topography of pollen grains. The examples shown here are from lily (*Lilium longiflorum*).

Plate 12a Pollen grains are seen lying in an anther loculus at the time of dehiscence. The remains of the tapetum and the loculus wall lie along the bottom of the picture. The organization of the pollen grain wall is determined at a very early stage of development. In lily pollen there are two distinct regions of wall. The smooth, unpatterned area, the colpus, forms on the outward-facing wall of each grain in a tetrad. The remainder of the wall is sculptured in a polygonal pattern, described further below. The colpus seems to be the spatial and functional equivalent of the pore seen in oat pollen in Plate 11. Its exine is relatively thin, and several examples showing its inward collapse are included. There is also a very small and shrivelled sterile grain. × 640.

Plate 12b This view shows part of a pollen grain (upper half of micrograph) lying on a carpet of orbicules (more or less spherical bodies, coated with sporopollenin, seen in the lower half of the micrograph). The layer of orbicules may aid the dispersal of the pollen when the anther loculus matures and splits open. Their appearance in ultra-thin section is shown in Plates 11b, c and 22. × 4500.

Plate 12c The architecture of the patterned part of the pollen grain wall is seen here at × 13500. The major difference between lily and oat (Plate 11) exines is that, in the former, the columnar bacula are not crowded all over the surface, but are restricted to the sides of irregular polygons. The tectum, or 'roofing layer', on top of the bacula, of necessity lies in the same polygonal pattern. In the central regions of the polygonal areas one can see down to the relatively smooth outer surface of the nexine. It is remarkable that whilst the position of the colpus is determined in relation to the cell axes in the tetrad, the genes that govern the detailed pattern of bacula and tectum exert their influence earlier, in the pollen mother cell, that is at a stage similar to that in Plate 11a, prior to the appearance of either intine or exine. Once the genes have been transcribed, the information they contain is present in the cytoplasm, ready to be utilized at the appropriate time during pollen maturation. Thus experimentally produced fragments of cytoplasm from mother cells or young pollen will develop the patterned exine even if nuclei are absent.

The architecture shown in these micrographs is characteristic of lily pollen, and similarly, pollen grains of other species develop their own specific shapes and patterns. Some of the wide range of possibilities should be evident from the differences observable between oat (Plate 11) and lily (Plate 12).

(We thank Drs. J. Heslop-Harrison and H. Dickinson for these micrographs, reproduced by permission from: 12a, *27th Symposium Soc. Dev. Biol. Suppl.* 2, 1968, Academic Press; 12b, *Planta*, **84**, 199-214, 1969, Springer-Verlag; 12c, *Society for Experimental Biology Symposium*, **25**, 1971).

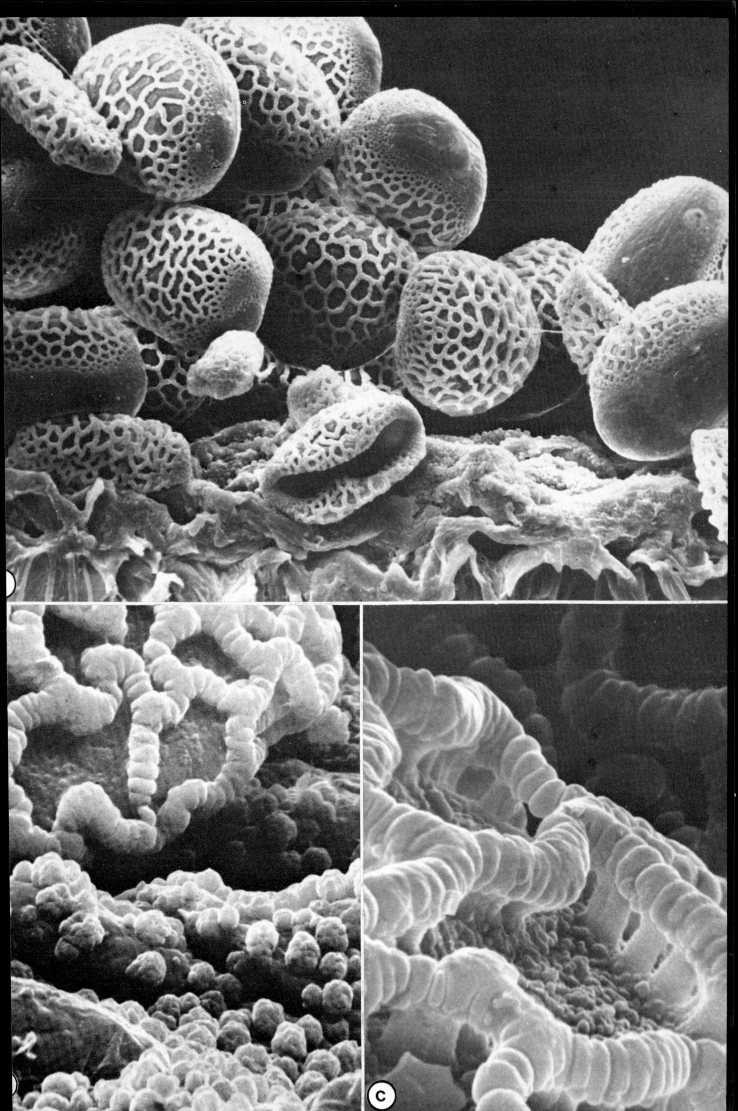

Plate 13

Transfer Cells

In transfer cells, the surface area of the plasma membrane is augmented by the development of ingrowths of wall material which protrude into the living contents of the cell. 13a illustrates their appearance in the light microscope and 13b in the electron microscope. The functional significance is very different in the two cases selected for inclusion in this Plate.

Plate 13a Transverse sections of vascular bundles in the root nodules of nitrogen-fixing leguminous plants show that each bundle is surrounded by a layer of endodermal cells, identified in the picture by the circlet of arrowheads pointing to the Casparian strips (see also Plate 16). Within the endodermis the pericycle cells nearly all have a fuzz of darkly-stained cell wall ingrowths (e.g. circles). Ingrowths are not prominent near the phloem (P) but do back on to the xylem (X). There is a complex two-way traffic of solutes through these cells. Sucrose enters the nodule *via* the phloem and passes through plasmodesmata to the cells (not shown) where nitrogen-fixation occurs. It is utilized in respiration and to provide carbon skeletons for the manufacture of nitrogenous amino acids and amides. The latter diffuse through plasmodesmata back to the vascular bundles. Within the endodermis the pericycle transfer cells are thought to secrete the nitrogenous solutes into the extra-cytoplasmic spaces, mainly cell walls, from where they diffuse into the xylem elements, and are flushed out of the nodule towards the root to which it is attached, by osmotic uptake of water. Sweet pea (*Lathyrus*) nodule, × 600.

The legume nodule is, in effect, a gland with an internal duct (the xylem). The secretory activities are carried out by the pericycle transfer cells, presumably aided by the large surface area of plasma membrane that they possess.

Plate 13b In this example, the transfer cells are xylem parenchyma cells modified by the development of wall ingrowths. They surround two xylem elements in a vascular bundle, sectioned just at the level at which it leaves the stem and passes out into a leaf. Such 'leaf traces' often display spectacular arrays of transfer cells. By using radioactive tracers it has been found that they can absorb solutes from the xylem sap, but it may be the case that they can also load it with solutes. It appears that departing leaf traces are for some reason especially active zones for these exchanges between the xylem and the surrounding living cells. Parts of the xylem other than at departing leaf traces (e.g. Plate 6d) have xylem parenchyma cells unmodified as transfer cells.

The way in which the wall ingrowths enhance the surface area of the plasma membrane (PM) is clearly visible. The ingrowths can branch and anastomose. Because they are finger-like projections from the wall and bend in all directions, they often appear in the section as apparently isolated profiles. Transfer cells of all types characteristically have many mitochondria (e.g. M in cell at upper right).

The two xylem elements themselves display several of the features also seen in Plate 6—microfibrillar remnants of the primary wall (arrowed circle), survival of microfibrils plus matrix in zones protected by lignin (arrows), and lignified thickenings (seen both in transverse and longitudinal section here because the xylem elements have been sectioned obliquely). *Galium aparine* seedling, × 12 000.

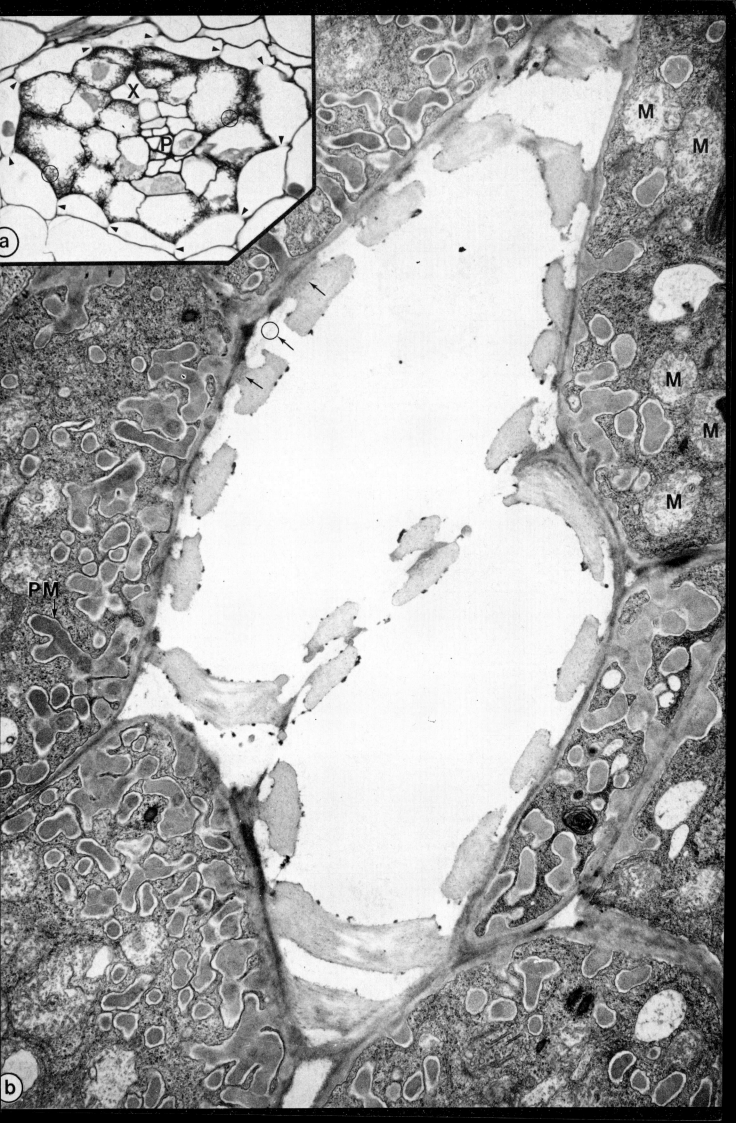

Plate 14

Plasmodesmata

This collection of micrographs illustrates several aspects of plasmodesmatal ultrastructure. Plasmodesmatal canals are seen 'end on' (14b, d) as well as in side views (14a, c, f–h). When interpreting the latter it is well to remember that the diameter of the canals is not very different from the thickness of the sections. Also the canals can pass out of the thickness of the section and so apparently disappear. Their appearance depends greatly on whether they lie symmetrically *within* a section or whether the knife edge slices through them so that parts of an individual are present in two adjacent sections.

Plate 14a Three plasmodesmata are shown here piercing the wall between two meristematic cells in a cabbage root. Small arrowheads follow the path of the plasma membrane from cell to cell in the lowermost example. In each plasmodesma there is a narrow axial strand (large arrowheads) almost certainly continuous with cytoplasmic cisternae of endoplasmic reticulum (followed by the small arrowheads in the topmost example). Such strands have been called desmotubules. × 45 000.

Plate 14b End-on views of very numerous plasmodesmata in a section that grazes a curved cell wall in the stele of an *Azolla* root meristem. The magnification is × 12 000, so a convenient way of counting the number of plasmodesmata per unit area is to cut a 12 × 6 mm window in a piece of paper, lay it over different areas, and obtain an average count. The true area of the window is 1.0×0.5 μm. The same method can be applied to the nuclear envelope pores (e.g. white arrows). The plasmodesmata tend to occur in rows (e.g. the row of five between the arrowheads). N, nucleus.

Plate 14c, d and e Plasmodesmata are often grouped in primary pit fields.

In 14c (oat leaf, × 31 000) the group is at a point of contact between mesophyll cells; 14d (pea leaf, × 66 000) is a similar situation, seen in oblique/surface view; 14e (root tip of maize, × 16 000) is a shadow-cast preparation showing the cellulose microfibrils delimiting the pit area and the individual plasmodesmatal pores. Microfibril patterns can also be discerned in 14d, an image which can be compared with that in Plate 8e.

14c and d both show the presence of axial desmotubules (arrows) in the plasmodesmatal lumina. In 14c the plasma membrane is seen to be closely constricted around the desmotubules at the cytoplasmic extremities of the plasmodesmata (e.g. circles). Note also the strange convolutions (arrowheads) in the central part of these plasmodesmata.

Plate 14f, g and h These three micrographs illustrate compound plasmodesmata, in which *several* canals meet in the interior of the wall. In 14f (between two transfer cells in the phloem parenchyma of a lupin leaf vein, × 35 000) many canals radiate in both directions from the centre of the wall. Because they radiate at a variety of angles, serial sections would be needed to demonstrate continuity of cytoplasm between the two cells. In 14g (vascular parenchyma of *Polemonium* stem, × 95 000) a side passage interconnects two plasmodesmata. The dark-light-dark configuration of the plasma membrane is shown to be retained throughout. In 14h the upper, empty-looking cell is a sieve element and the lower its companion cell. Many plasmodesmata funnel from the latter towards fewer canals leading into the former. On the sieve element side the canals are lined with electron-lucent callose (*Lupinus*, × 35 000). In all three examples the wall is swollen at the site of the compound plasmodesmata. These swellings can easily be seen with the light microscope, especially those on the walls of sieve elements.

(We thank Dr. K. Muhlethaler and Elsevier Publishing Co. for permission to reproduce Plate 14e from *Ultrastructural Plant Cytology*, 1965).

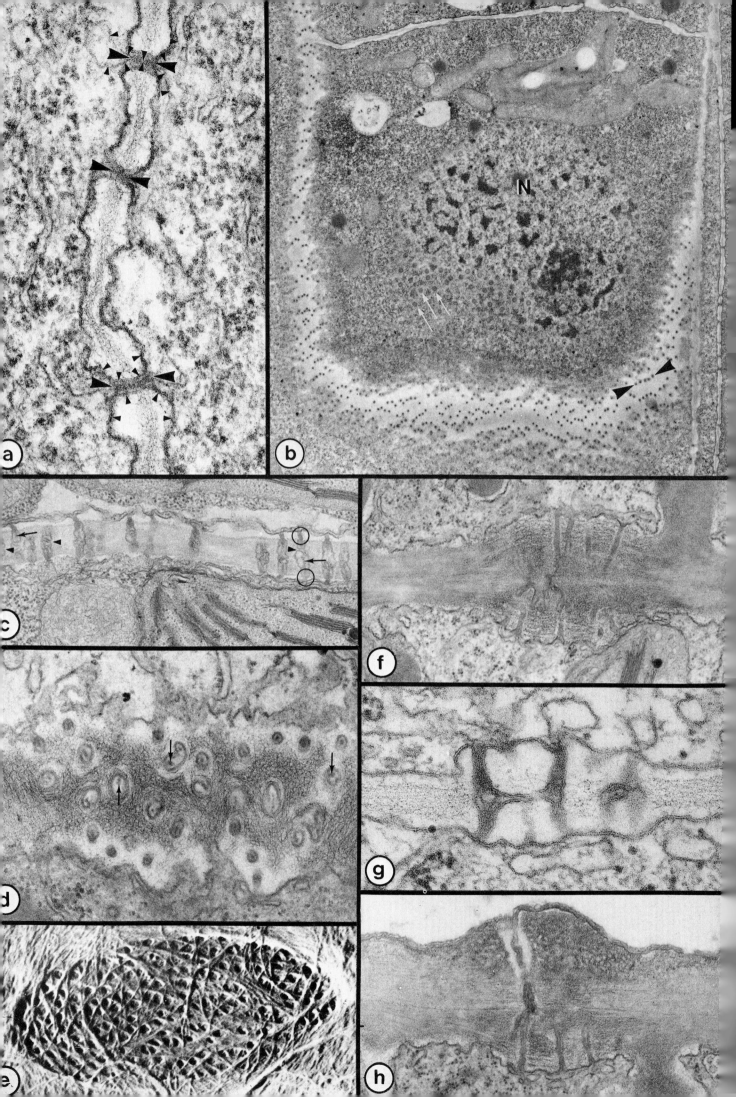

Plate 15

Pits

This collection of micrographs depicts two very different types of pit connection between cells that have massive secondary walls. One type (15a, b) is 'simple', and in fibres that remain alive for a considerable period; the other (15c-e) is 'bordered', and in tracheids that die at maturity. The former functions in movement of substances between living cells, the latter in movement of xylem sap in wood.

Plate 15a and b The two fibres (top and bottom halves of the picture) of 15a are still relatively young. The close approach of the plasma membrane to the middle lamella (ML) in the pit itself shows how thin the primary walls of the two neighbours originally were. Subsequently the cells deposited very thick secondary walls (S) everywhere *except* in the immediate vicinity of the plasmodesmatal connections (PD) in the pit regions. The secondary walls are microfibrillar at this stage, and not yet impregnated with lignin. The more they become lignified, the more important the symplastic connections are if the cells are to remain alive. A full complement of cytoplasmic components is present—chloroplast (C), mitochondrion (M), dictyosomes and associated vesicles (D), and endoplasmic reticulum (ER).

The layer of secondary wall does not overarch the pit, as shown both in the section and in the scanning electron microscope view (15b). The pit is therefore described as *simple*. Plate 15a, pea leaf, × 48 000; Plate 15b, oat leaf, × 7000.

Plate 15c, d and e These scanning electron micrographs show structures in the side walls of xylem tracheids in a piece of pine wood. The wood was split open longitudinally to expose the insides of the cells to view. The pits occur in rows, and occupy a large fraction of the side walls. The diameter of the tracheid in Plate 15c is approximately the distance between the two black sloping lines. Three pits are included in this image. That on the left has retained its cowl, or border, of secondary wall; by chance the other two have had the border split off. Structural components lying within the border are seen at higher magnification in 15e. When they in turn are removed the inner face of the border on the opposite side of the pit is revealed (15d). Thus these bordered pits consist of two opposed cowls, with a central *pit membrane* between them.

Whereas the pit 'membrane' (it is a wall structure and not a true cell membrane) is at the level of the middle lamella and primary wall, the overarching borders are composed of heavily lignified secondary wall. The pit membrane consists of a lens shaped disc of lignified wall material (the *torus* (T)) suspended in the cavity between the pair of borders on bundles of microfibrils largely arranged like the spokes of a wheel (MF). The diamter of the torus is somewhat greater than that of the apertures of the borders. It can move on its slings of microfibrils so that it is either free in the centre, or it can be pushed sideways to lie snugly against one or other of the apertures. In other words it acts as a valve. When the interconnected tracheids are both full of columns of sap ascending the tree the valve will be in the central, open position. Lateral transfer through the large gaps in the microfibrillar sling is possible. If the negative tensions that 'pull' the sap up the tree lead to cavitation and air bubble formation in one tracheid, the change in pressure will cause the valve to swing across and seal off the damaged element.

Formation of the pit membrane (torus plus sling) during the maturation of tracheids involves phenomena similar to those described for maturation of primary xylem elements. The primary wall is hydrolysed *except* where it is protected by lignin, to leave only matrix-free cellulose—in this case the microfibrils of the sling. The warty excrescences seen through the sling in 15e and in full view in 15d also arise in the final stages of maturation. They may derive from remnants of cytoplasm that come to lie scattered over the inner face of the wall.

Plate 15c, × 1500; 15d, × 3400; 15e, × 7600.

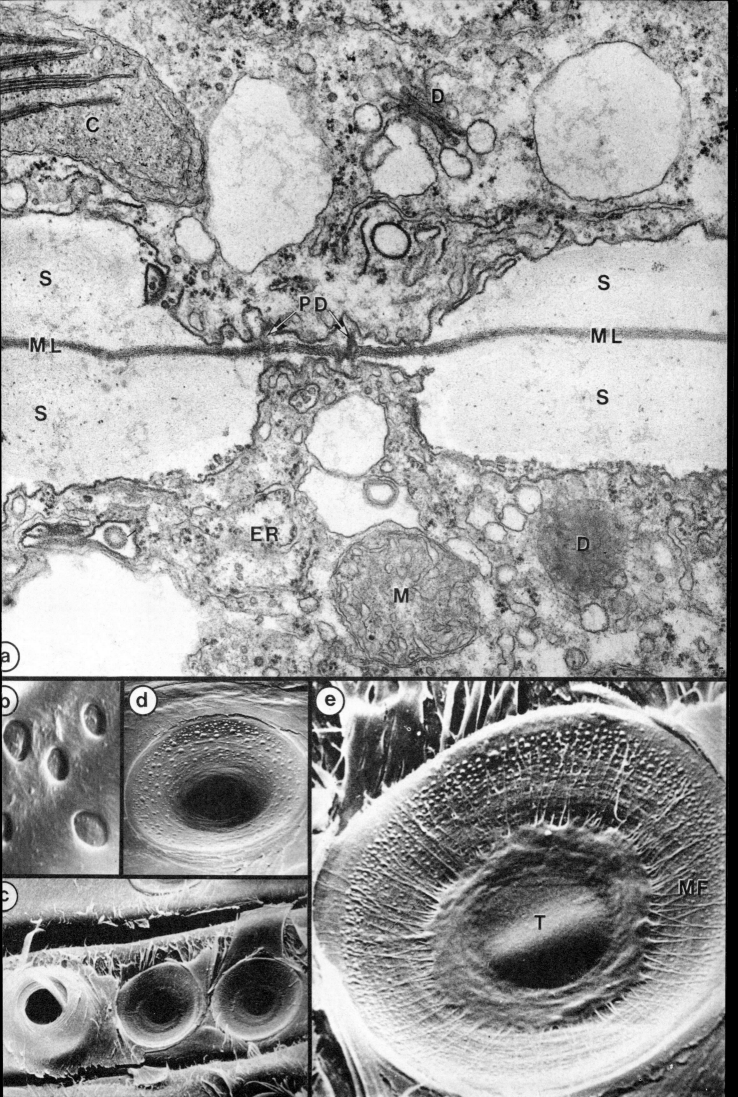

Plate 16

Endodermis and Casparian Strip

The location of the endodermal cell layer in a root is shown in a light micrograph, and the two electron micrographs then focus on the structure of the Casparian strip in the radial walls of the endodermal cells.

Plate 16a Transverse section of the central stele (vacuolar tissue) of an *Azolla* root. The cell types are labelled so that reference can be made to Plate 49, which shows the same types in the same spatial arrangement in an immature part of the root. E, endodermal cell; P, pericycle cell; S, sieve element; PP, phloem parenchyma; PX, protoxylem; MX, metaxylem. Arrowheads point to the Casparian strips, which were differentially stained by the technique used. The lowermost is the most clearcut example. × 1700.

Plate 16b The radial wall separating two endodermal cells (E) runs across the bottom of this micrograph. Part of a cortex cell is seen at the left, and part of a pericycle transfer cell at the right. The material is the same as that viewed in Plate 13a (a vascular bundle in a legume root nodule), indeed the orientation of the cells is much as at the arrowhead on the extreme left of that picture.

The Casparian strip lies between the open arrows. Even at this low magnification it can be seen to have a more homogeneous texture than the neighbouring wall. For instance the middle lamella (ML) almost disappears in the strip. Notice too that the undulations of the plasma membrane become smoothed out at the Casparian strip. × 15 000.

Plate 16c A portion of Casparian strip (top) is here contrasted with a portion of normal cell wall (bottom) that happened to lie in the same field of view. The smooth texture of the strip is presumably due to impregnation of all of the spaces between the microfibrils with cutin or suberin. The plasma membrane (PM) shows the dark-light-dark appearance on both walls, but very conspicuously so at the Casparian strip because it is unusually flat, and, in this case, lies at right angles to the plane of the section. It has been found that if endodermal cells are plasmolysed, the plasma membrane sticks tenaciously to the impregnated Casparian strip. The membrane can even rupture down its centre line, the cytoplasmic face pulling away with the cytoplasm and the wall face remaining attached to the strip—much as when frozen tissue is fractured as part of the freeze-etching process, where again membranes tend to rupture along their mid plane.

Other membranes are present. The narrow compartment across the middle of the picture is a flattened vacuole (V), bounded by its tonoplast (T). The thickness of the tonoplast more or less matches that of the plasma membrane, and by contrast the membrane of the endoplasmic reticulum (ER) is seen to be much thinner—its 'tramline' configuration is only just detectable (circle). Material as in 16b. × 120 000.

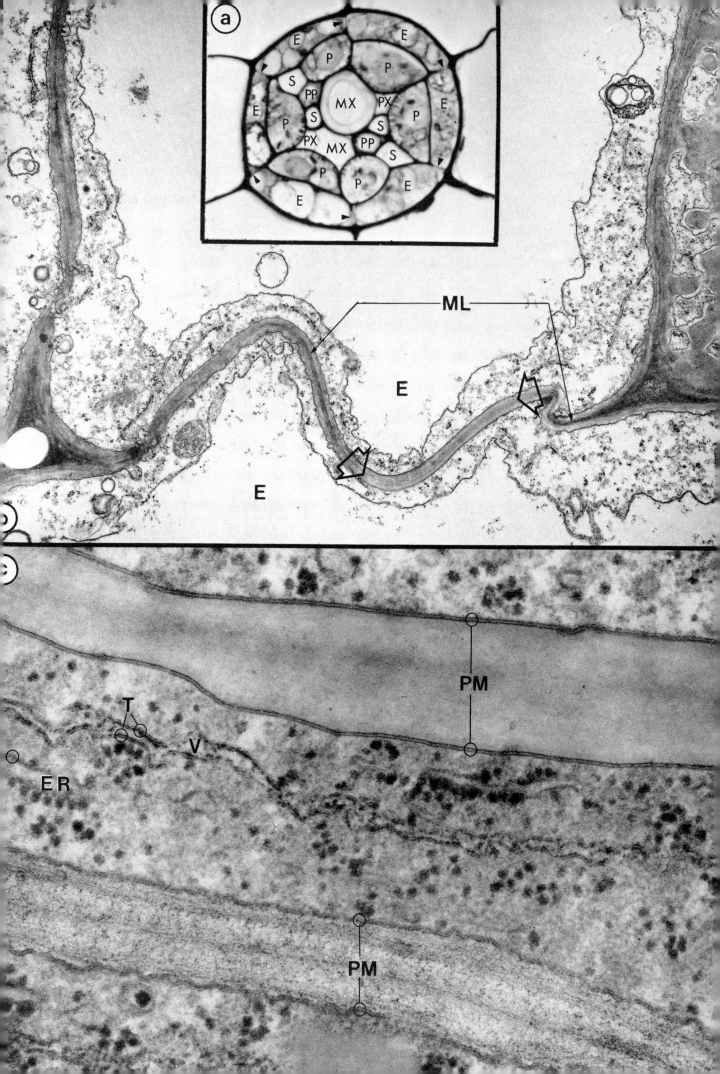

Plate 17

Vacuoles

Hardly any low magnification view of plant cells is without vacuoles. They are present in most of the micrographs in the book. The three pictures collected here illustrate specific features.

Plate 17a These views of the tonoplast were obtained by the freeze-etching technique. The fracture plane has passed *over* one vacuole (left) and *under* its neighbour (right), probably along the mid line of the tonoplast membrane. The resulting convex and concave surfaces are therefore the vacuolar (V) and cytoplasmic (C) halves of the tonoplast, viewed from its cleaved interior. The tonoplast is asymmetric. Particles lie scattered or in short rows in the convex vacuolar half, while in the concave cytoplasmic half they are so crowded that no smooth surface can be seen between them. It is not known whether the particles in the two faces are chemically similar or not, or whether proteinaceous solute pump molecules are included amongst them.

Parts of other cell components lie in the cytoplasm surrounding the vacuoles—notably cisternae of endoplasmic reticulum (ER).

Magnification × 60 000. (We thank Dr. B. A. Fineran and Academic Press for permission to reproduce this micrograph from *J. Ultrast. Res.*, **33**, 574-586, 1970).

Plate 17b The tonoplast (T) of this vacuole (V) in a soybean root tip cell passes around a complex invagination, connected to the cytoplasm at a narrow isthmus. The tripartite dark-light-dark substructure of the tonoplast is revealed, along with its asymmetry, the vacuolar face being darker and carrying a surface deposit.

Whorls of membranous material within the invagination also show dark-light-dark triple layering where their orientation with respect to the plane of the section is appropriate. In places it can be seen that a thin zone of non-membranous material is sandwiched between adjacent whorls, alternating with clear inter-membrane spaces. This inclusion could therefore have originated from a set of concentrically arranged cisternae by a process of autophagic digestion. The figure shows but one example of many irregular forms of membranous vacuolar inclusion. While most such inclusions can be interpreted in terms of digestion, it is unfortunate that they could also be artifacts produced by bad fixation. In the present case the remainder of the cell and tissue showed no signs of damage, so the inclusion is at least likely to be 'real', whether or not it depicts autophagic digestive activity. × 125 000.

Plate 17c Many of the compounds stored in vacuoles are of low molecular weight and leak out of the tissue when it is being prepared for electron microscopy. Some, however, are retained and appear in electron micrographs as non-membranous vacuolar inclusions.

The flocculent particles seen here in the vacuole of a potato leaf cell probably consist of a protein that inhibits protein-digesting enzymes (e.g. trypsin, chymotrypsin) of animals. Many organisms, from bacteria to flowering plants, produce this class of protein, and species in the family Solanaceae (like potato and tomato) accumulate one type, known as 'Inhibitor-1'. They do so particularly when the leaves are bruised or, as in the example shown, cut off the plant. It may be that Inhibitor-1 is one of the extremely numerous 'defence chemicals' of the plant kingdom, in this case conferring the biological advantage of reducing the palatability of the plant to herbivorous animals. Potato leaf × 18 000.

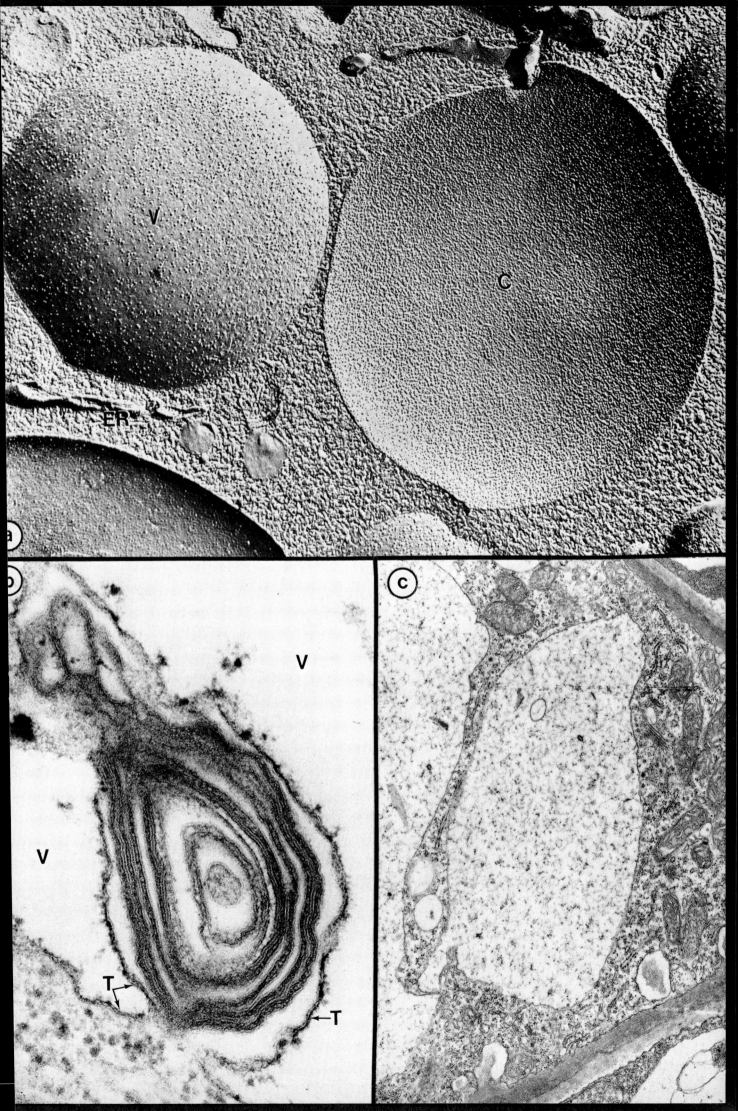

Plate 18

The Nuclear Envelope and its Pores

Plate 18a The fracture in this freeze etched preparation of the nuclear region of a *Selaginella kraussiana* cell has passed along the surface of the nuclear envelope, revealing the pores in surface view (lower half of picture), and has broken through the inner (I) and outer (O) membranes of the envelope. The upper half of the picture shows relatively featureless nucleoplasm (N). The pores are also seen in side view (arrows) at intervals along the cross-fractured nuclear envelope membranes. A continuity between endoplasmic reticulum (ER) and nuclear envelope may occur at the point marked by the star. The helical symmetry with which these nuclear pores are arranged is unusual. Magnification × 24 000. (We thank Drs. B. W. Thair and A. B. Wardrop for this micrograph, reproduced by permission from *Planta*, **100**, 1-17, 1971).

Plate 18b This three-dimensional reconstruction of part of the surface of a nucleus in a cress root tip cell was made by projecting the electron microscope images of nine adjacent sections on to sheets of polystyrene, to give a final magnification of × 150 000. The outline of the nuclear envelope in each section was cut out, the pores were marked, and the nine layers superimposed in the correct alignment. The scale represents 1 μm. The nuclear surface undulates, as indicated by the irregular 'contours' traced out by each layer of the model, and the pores lie in no obvious pattern, at a 'pore density' of 15-20 per μm^2 of nuclear envelope.

Plate 18c Where the nuclear surface is irregular, a single section can, as here in another cress root cell, include both side (S) and face views of pores in the nuclear envelope (NE). In some of the face views (arrows) a granule is seen in the centre of the pore. Although regions of heterochromatin (C) touch the inner membrane of the nuclear envelope, a halo of clear nucleoplasm lies around the pores (lower left). × 32 000.

Plate 18d Several details of pore structure can be seen in this glancing section of a nuclear envelope. The appearance of a pore depends on its position and orientation within the thickness of the section, which at this magnification (× 100 000) is a slice about 5 mm thick. The pore perimeter is octagonal, and the flat edges of the octagons are visible in several cases (pores 1, 3, 4, 8, 11). Other features shown include: particulate components of the inner annulus (identifiable by proximity to chromatin) (upper parts of pores 2 and 4) and outer annulus (identifiable by proximity to polyribosomes) (pores 5 and 9); fine fibrils traversing the pore lumen (pores 1, 7, 8, 10) and apparently nearly occluding some pores (pores 4, 9, 11); pores arranged equidistant from a clump of chomatin (C) (e.g. pores 6, 7, an un-numbered pore, 8, and 10); fibrils passing between chromatin and the pore margins (arrows); polyribosomes (P) on the outer surface of the nuclear envelope; particles in the centre of some (1, 2, 4, 6, 7, 8-11) but not all pores. *Vicia faba* root tip.

Plate 18e Side views of pore complexes are seen here, in the same material and at the same magnification as in 18d. The nucleoplasm is at the top of each picture. All examples show continuity of inner (I) and outer (O) envelope membranes at the pore margins, but other features vary. Part of the variation is due to the pores lying at slightly different levels with respect to the section thickness. Units of the annulus are seen at the inner margin (e.g. single arrows) and outer margin (e.g. double arrows). Particles thought to be preribosomal in nature (large open arrows) are: absent (top left); in the nucleoplasm near the pore (top centre); at the inner part of the pore lumen (top right); both inside and outside the pore (lower left); in the pore and just outside in the cytoplasm (lower centre); and in the pore as well as both inside and outside (lower right). These particles are smaller than cytoplasmic ribosomes, visible in the lower part of each micrograph. As in 18d, strands (chromatin?) are sometimes seen connected to the inner annulus.

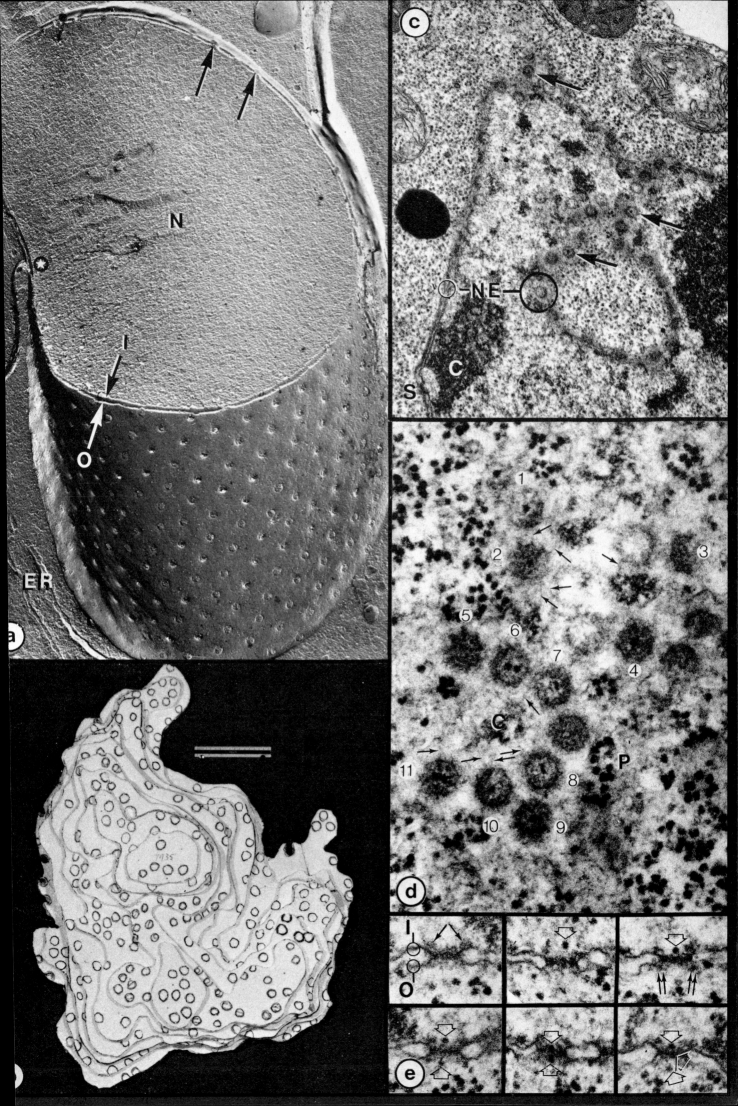

Plate 19

The Nucleolus

The nucleolus is a characteristic feature of the nuclei of eukaryotic cells. Major constituents are a repeated sequence of genes (DNA), and a mass of products (RNA) of the activity of those genes. Most of the RNA is a precursor of the RNA of ribosomes, and the dynamic processes observable in nucleoli centre on the synthesis of this material, first in fibrillar and later in granular form, and on the transport of the pre-ribosomal RNA from the nucleolus to the cytoplasm.

Plate 19a-f These six photomicrographs illustrate dynamic processes in a nucleus of a cell growing in artificial culture, the cultured tissue having been derived originally from a tobacco plant. The nucleolus (centre of each picture) is surrounded by a bright halo, due to the phase-contrast optical system used. At the start of the sequence (19a) the nucleus (margin outlined by arrows) is ellipsoidal. The nucleolus, about 7 μm in diameter, contains a large nucleolar 'vacuole' about 70 μm^3 in volume. One minute later (19b) the 'vacuole' is connected to (arrow), and apparently emptying into, the nucleoplasm. After another 15 seconds (19c) the 'vacuole' is much smaller. It is just detectable after a further 15 seconds (19d), but cannot be seen (19e) 2 minutes after the first picture was taken. By this time the whole nucleolus has shrunk by about the volume of the 'vacuole'. One hour later (19f) another 'vacuole' has formed in the nucleolus, which has regained its original total volume, and the nucleus itself has become more rounded. All × 1400. (We thank Dr. J. M. Johnson for these micrographs, reproduced by permission from *Amer. J. Bot.*, **54**, 189-198, 1967).

Plate 19g Part of a nucleolus in a *Vicia faba* root tip cell nucleus is shown here magnified × 58 000. A central 'vacuole' (V) is enclosed within the nucleolus (as in 19a, f). Granular (G) and fibrillar (F) zones constitute the dense material, along with small areas of chromatin (arrows) ramifying through electron-transparent channels. Condensed (CC) and dispersed (DC) chromatin is seen in the nucleoplasm; the former at one point touching and penetrating into the nucleolus (dashed lines). The 'vacuole' contains scattered fibrils and particles, and it is presumably they that are discharged into the nucleoplasm when 'vacuoles' pulsate.

Plate 19h This high magnification (× 150 000) view shows 6-8 nm nucleolar fibrils (F) and 12-14 nm granules (G) near the periphery of a nucleolus (*Vicia faba* root tip). The granules (see also Plate 18e) are smaller than cytoplasmic ribosomes. Some granules are attached to, or associated with, fibrils in a manner suggesting that they might be formed by folding and condensation of fibrils. Convoluted fibrils of DNA-histone are seen in the chromatin (C), near which lies a perichromatin granule (arrow), with angular profile, 50-60 nm in diameter.

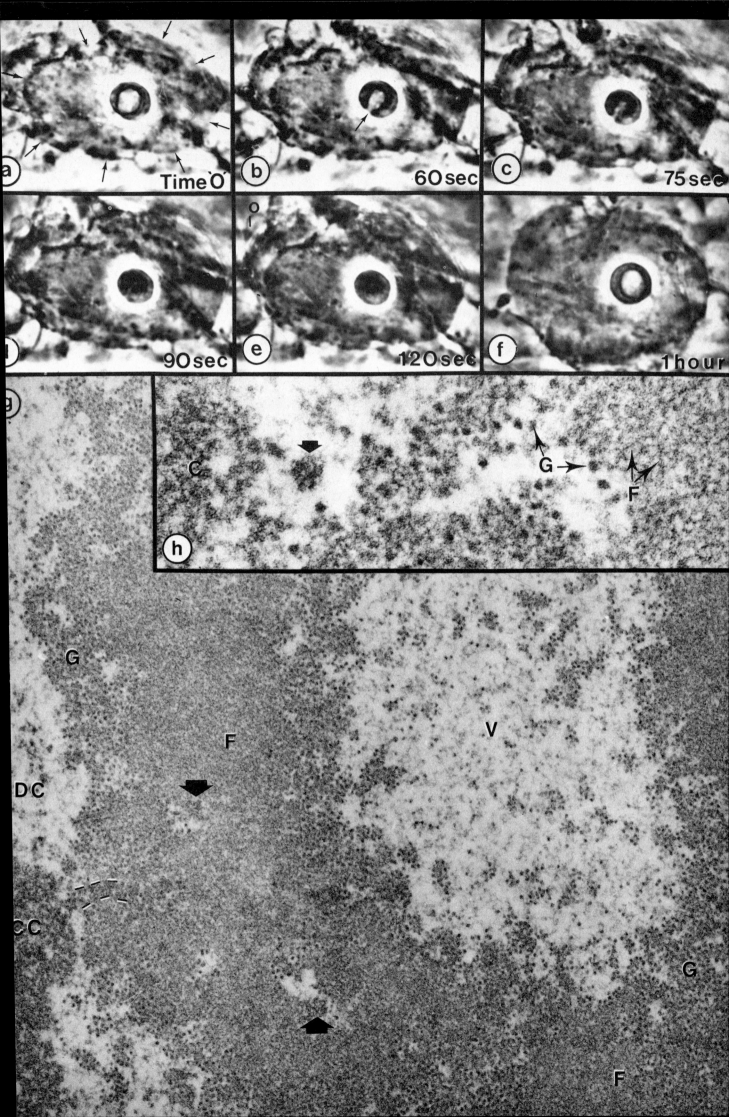

Plate 20

Plates 20-24 depict a range of specialized plant cells, and are grouped together in order to illustrate the diversity of structure and function exhibited by the endoplasmic reticulum.

The Endoplasmic Reticulum, Polyribosomes, and Protein Synthesis in Cotyledon Cells

The cotyledons of legume seeds manufacture and store large amounts of protein. Developing and nearly mature cells are shown here.

Plate 20a This light micrograph shows a typical cotyledon cell at an advanced stage of development, stained by the periodic acid—Schiff's reagent procedure (which reacts with carbohydrates, as in the cell wall and in starch grains) and with bromophenol blue (which stains basic materials such as proteins). The central nucleus (N), containing chromatin and a nucleolus with 'vacuoles', is surrounded by large starch grains (S) (stained pink in the original preparation, appearing dark grey here with a lighter central region) and protein bodies (PB) of various sizes (stained dark blue in the original, black here). These are interspersed with a vacuolated cytoplasm which contains an extensive endoplasmic reticulum system (open arrows). Note also the intercellular spaces and the darkly stained material (probably proteinaceous) deposited in the corners of the walls lining these spaces. *Lathyrus* (sweet pea) seed, × 660.

Plate 20b Cotyledon cells at an early stage of development contain an extensive endoplasmic reticulum, the cisternal membranes of which are covered with spiral polyribosomes. These ribosomes assemble the storage proteins of the cotyledon, principally *vicilin* and *legumin*. As the length of a polyribosome spiral corresponds approximately to the length of the messenger RNA molecule that is being translated, it is possible to estimate the molecular weight of the equivalent protein. The longest polyribosomes in this micrograph contain 20 ribosomes in a total spiral length of about 500 nm. Three nucleotides are required to code for one amino acid, and they occupy a length of approximately 1.0 nm of the RNA strand, implying a possible total of about 500 amino acids, and if each has an average molecular weight of 120, the total molecular weight of the protein is estimated to be about 60 000. In fact, biochemical analyses have shown that legumin and vicilin are complex proteins, each made up of several amino acid chains, the largest of which have molecular weights within 10% of 60 000. Magnification × 26 000 *Vicia faba*, (the broad bean), young cotyledon.

Plate 20c As the protein accumulates it is transferred by an undetermined route to large protein bodies (PB) from the endoplasmic reticulum (ER). Individual protein bodies have been shown to contain both vicilin and legumin molecules. It is thought that protein bodies arise by the accumulation of protein within vacuoles. In addition, lipid stores begin to accumulate in the form of droplets (small arrows), frequently associated with the endoplasmic reticulum, but external to the cisternae (e.g. beside the asterisk). Note also the dictyosome (D) which has become so reduced as to be scarcely recognizable. In this micrograph and the previous one, the paucity of free ribosomes is evident. (*Vicia faba*, nearly mature cotyledon, × 14 000).

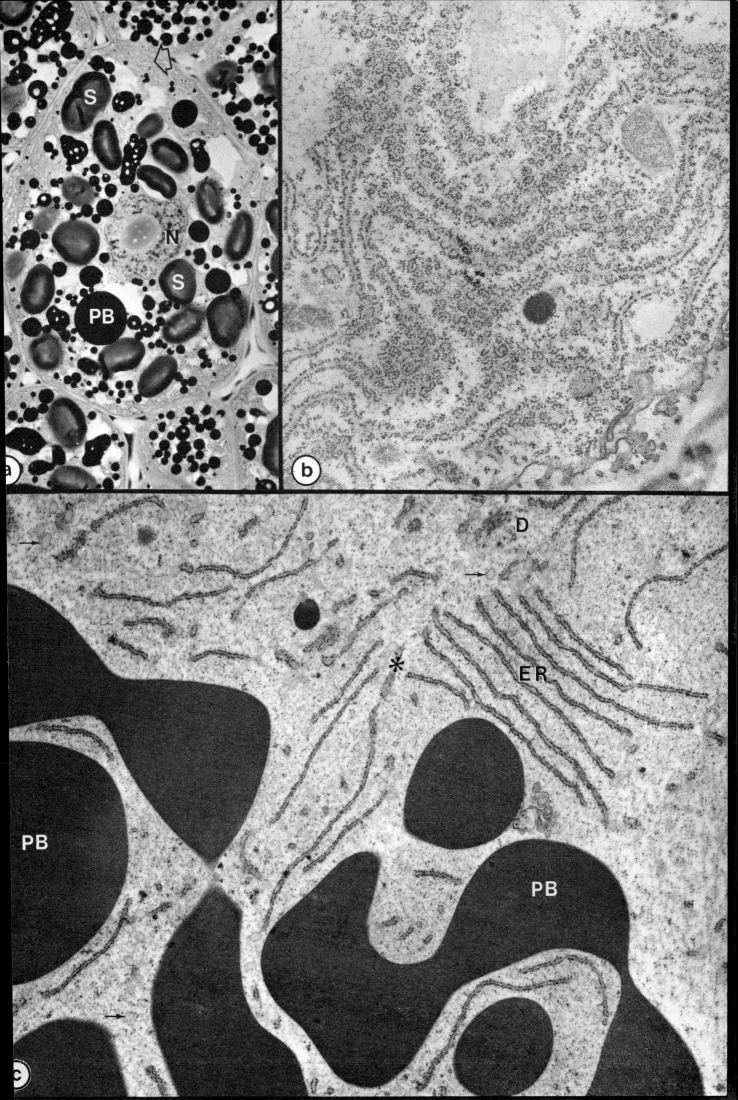

Plate 21

The Endoplasmic Reticulum and Polyribosomes

Plate 21a This unusually regular arrangement of the endoplasmic reticulum is found in surface glands (trichomes) on leaves of *Coleus blumei*. The 12 to 14 parallel cisternae are stacked so closely that the intervening cytoplasm is restricted to very thin layers, which, however, expand in some places (stars). The cisternae are interconnected at branch points (arrows) and in a few places by swollen intracisternal spaces (S), which contain diffusely flocculent material, perhaps the product synthesized by the system. The product (Pr) secreted by the gland passes through the inner layer of the cell wall (CW) and accumulates beneath the cuticle (C). The endoplasmic reticulum is continuous with the outer membrane of the nuclear envelope (open arrow, lower right). The stroma of the plastids (P) is filled with a material which (after specimen preparation) is very dense to electrons. The gland probably secretes mono- or di-terpenes. × 22 000.

Plate 21b One of the spaces into which the cisternae open is shown here enlarged from 21a. The membranes of the endoplasmic reticulum have a just discernible tripartite substructure (black and white arrow) and are thinner than the adjacent tripartite plasma membrane (black arrow).

It is clear that the intra-cisternal cavity (Ci) is not in open communication with the space outside the plasma membrane (Co), but several clues in the micrograph suggest that this arrangement of membranes may alter with time, and that cisternal contents can be released to the exterior. The clues are: similarity in size and flocculent contents between Ci and Co, suggesting that the former could be a reservoir which is filled up and discharged to the exterior; the presence of fragments of membrane in Co, as if they had been cast off during the previous discharge; one fragment (open arrow) not only has a tripartite construction of the same dimensions as the plasma membrane, but also encloses several ribosomes, perhaps derived from the narrow bridge of cytoplasm that separates Ci and Co. It should be noted that although the endoplasmic reticulum and plasma membrane are in close proximity over much of the periphery of the cell (21a), there is no evidence that the two can fuse. Rather, presence of fragments of membrane in Co suggest that discharge of Ci may be by explosive rupture of the membranes following increase in internal pressures, rather than by fusion of endoplasmic reticulum and plasma membrane. On this hypothesis, the two classes of membrane would heal their ruptures after the pressure had been released, again without fusion of unlike membranes. These speculations illustrate the way in which a static image can be used to hypothesize about dynamic processes; it is, however, important to recognize that to confirm or reject the ideas requires further experiments. × 75 000.

Plate 21c Two polyribosome conformations which differ from the spirals seen in Plate 20b are shown here. In the main picture, which shows oblique surface views of cisternae of rough endoplasmic reticulum in a hair cell of the alga *Bulbochaete*, the polyribosomes often lie in parallel chains. Polyribosomes are also seen on the outer surface of the nuclear envelope (N-nucleus). × 67 000. (We thank T. Fraser for this micrograph).

The insert shows a more common configuration, which occurs free in the cytoplasm, unattached to membranes. It consists of a tight helix of ribosomes. The example shown is from a bean root tip cell, and because the helix is much longer than the section is thick, there is no guarantee that the 0.4 μm length included here represents the total length. Comparable polyribosomes in liver cells have been shown to be based on unusually long strands of helically wound messenger RNA. × 90 000.

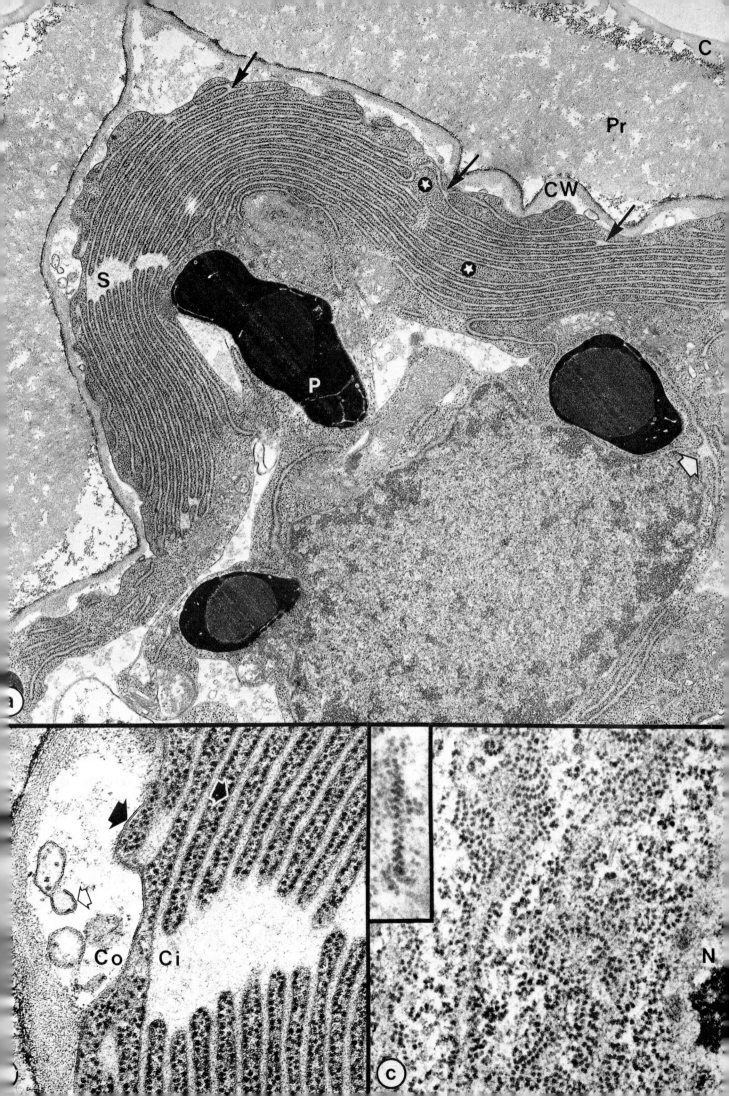

Plate 22

The Cytoplasm of Tapetal Cells

Cells of the tapetum have already been introduced in Plate 11, and further details of their cytoplasm are shown here using as an example an anther of *Avena strigosa*, containing developing pollen grains.

Plate 22a–c These cells are most unusual in that they lack vacuoles and (at maturity) cell walls. In several respects they resemble an animal epithelium, with one face (on the right in (a)) directed towards the anther locule (LO), and the other (on the left) based on the cells comprising the anther wall (see also Plate 11). A small portion of a pollen grain intrudes at upper right.

Orbicules (O) and remnants of cell wall material (arrows) (see Plates 11 and 12) are present on the locular side of the tapetal plasma membrane (PM). The cytoplasm is filled with plastids (P), mitochondria (M), dictyosomes (D), and both rough and smooth endoplasmic reticulum (ER).

The rough endoplasmic reticulum tends to lie in aggregates of fairly parallel cisternae (as in liver cells). Its polyribosomes are spiral. The rough cisternae are continuous with smooth regions, as shown at the dashed lines in (a) and (b). Both types of cisternae contain a material of moderate electron density, but the smooth parts tend to be dilated into sausage-shapes or spheres (22c, arrowhead) except in the case of smooth cisternae that ensheath plastids and mitochondria (best seen in 22b—arrowheads).

The cytoplasm contains numerous large vesicles (V) with flocculent contents. Other micrographs suggest that these are derived from the dictyosomes, and profiles such as the one marked by an asterisk (lower left) that they liberate their contents at the plasma membrane by reverse pinocytosis (not necessarily at any one face of the cell).

The cell wall structure seen on the left hand side of 22a is complex. The plasma membrane (PM) is not in close contact with the wall, which probably consists of an electron-transparent lipoidal deposit (L) on an otherwise conventional type of wall (open arrow). Some plasmodesmata (PD) pierce both layers. The composite wall may be a seal that reduces outward losses of nutrients leaking from the locule between the tapetal cells.

Provision of nutrients for developing pollen is known to be a major function of the tapetum. It also synthesizes certain proteins which become incorporated into the exine of the mature pollen, and it manufactures the cores of orbicules which lie on the locular surface, where, like the pollen grains themselves, they become coated with sporopollenin. The sub-cellular source of the precursors (probably carotenoids) of the sporopollenin has not been identified with certainty: a, × 28 000; b, × 48 000; c, × 48 000.

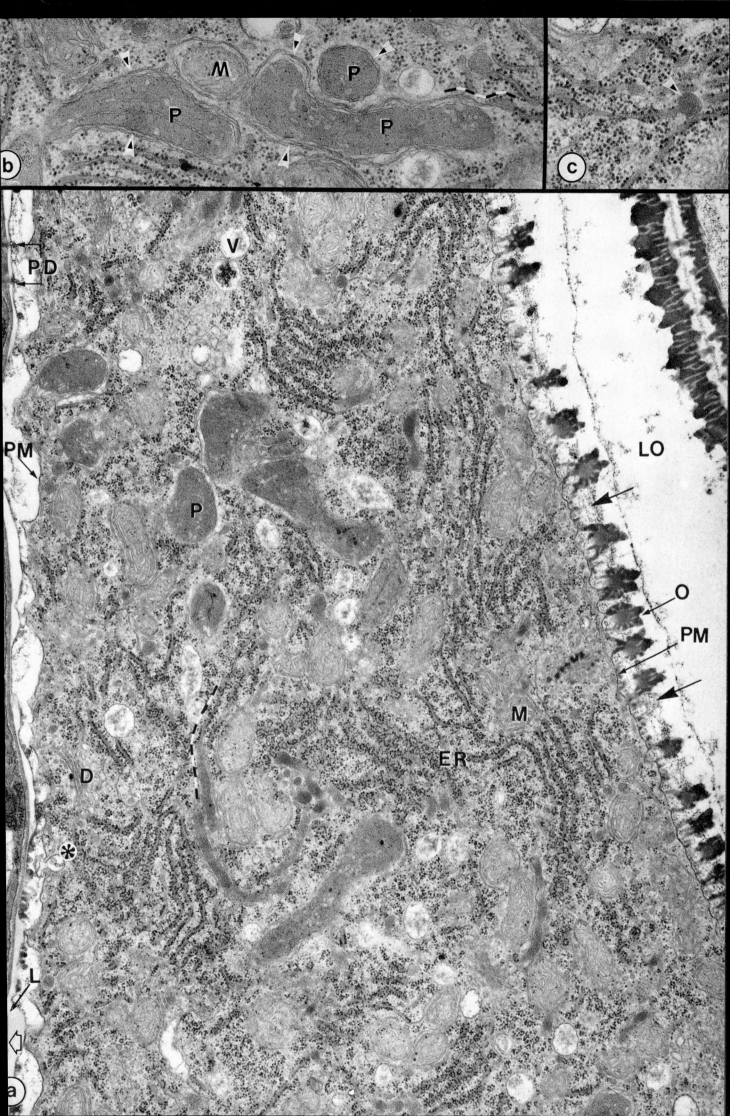

Plate 23

Smooth Endoplasmic Reticulum in 'Farina' Glands

Plants manufacture and secrete an enormous diversity of chemical substances. The example illustrated here, a white floury material, or 'farina', is produced by single celled glands developed particularly by members of the flowering plant family *Primulaceae*. Farina glands are one of several categories of plant gland in which a system of smooth endoplasmic reticulum membranes is exceptionally well developed.

Plate 23a and b These scanning electron micrographs show the floury appearance of the surface of an *Auricula* sepal, visible to the eye as a yellow powder, and at low magnifications ((a), × 1100) as mounds of fine crystals. Each mound ((b), × 5800) is a gland cell, covered by its secretion product in the form of contorted ribbon-shaped or linear crystals radiating from the cell surface. A mixture of substances is present, the major components being flavonoids. Some people are allergic to the 'flour'.

Plate 23c In the course of preparation for ultra-thin sectioning, this material (a farina gland on a young petal of *Primula kewensis*, × 40 000) had its secretion product dissolved away. Consequently none is seen outside the cell wall (CW). The main feature of the cytoplasm is the system of smooth endoplasmic reticulum tubules, each one 60-100 nm in diameter, ramifying through the cytoplasm. Not many branched tubules are visible (arrows), and it is obvious from the varied profiles displayed in the section that the tubules do not lie in a regular pattern. The ground substance of the cytoplasm is granular, and contains a rather sparse population of ribosomes, usually in clusters, presumably polyribosomes (e.g. circle). The mitochondria (M), being nearly circular in outline, are probably nearly spherical, but the electron-transparent areas within them suggest that they might have become swollen during preparation. Microbodies (MB), with granular contents and single bounding membrane, are conspicuous. The plastids (P) are small and simple. An electron-dense deposit lines parts of the tonoplast of several of the vacuoles (V). In a number of places (asterisks) the configuration of the tonoplast suggests that material such as the small droplets of electron-dense material that are present throughout the cytoplasm (especially at lower left) associated with the endoplasmic reticulum, may be being incorporated into the vacuoles.

The significance of this cytoplasmic organization in relation to the synthetic and secretory activity of the farina gland is obscure. The plasma membrane (PM) is fairly smooth, and there is no evidence that products of secretion leave the cytoplasm in the form of vesicles by reverse pinocytosis (*granulocrine* secretion). It has been suggested that the outward transport is *eccrine*, that is, by a flux of free molecules (not in vesicles) across the plasma membrane, and thence to the exterior by diffusion through the cell wall. Crystallization occurs upon exposure to the air. The labyrinths of smooth endoplasmic reticulum are thought to be associated with the synthesis of terpenoid substances, for they are present in a variety of plant glands secreting fats, oils, and fragrant essential oils. Flavonoid molecules are in part derived from the same precursors as terpenoids.

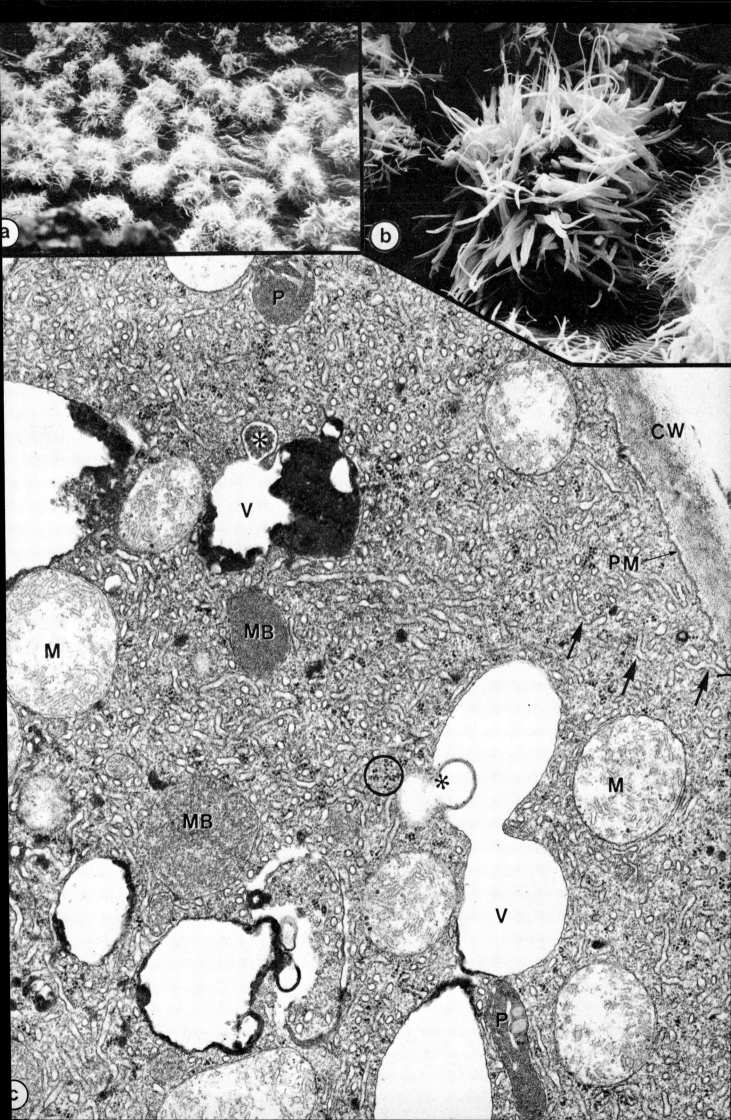

Plate 24

Developmental Changes in the Endoplasmic Reticulum of Sieve Elements

The endoplasmic reticulum can alter greatly during maturation of a cell. This plate illustrates one such developmental sequence, taking as an example the specialized endoplasmic reticulum found in sieve elements (the 'sieve element reticulum'). Plate 8b shows the reticulum (ER) in a mature sieve element, and should be examined before the more detailed pictures presented here.

Plate 24a Young sieve elements possess a full complement of cell components, though few are represented in this transverse section (*Coleus blumei* stem apex). The tonoplast (T) delimits the vacuole (V) at this early stage of development, but is destined to break down, so that cytoplasmic and vacuolar material will intermingle as what has been called *mictoplasm*. Irregular cisternae of rough endoplasmic reticulum (ER) are present in quantity throughout the cytoplasm, and remain while the cell synthesizes masses of a proteinaceous material (P), formerly described as 'slime bodies' and now as P-protein bodies (P being an abbreviation for phloem). An unusual type of vesicle is found in these cells, but is not restricted to them. It is spherical (see circled example in insert), about 50 nm in diameter, and spines, from which the name *spiny vesicle* is derived, radiate from its surface. *Coated vesicles* (e.g. Plates 27 and 28) are similar, but with a less distinctive coat. Spiny vesicles are usually present in clusters, and in this micrograph they lie (arrows) between massed endoplasmic reticulum cisternae and a growing P-protein body. It is conjectural whether they have a role in transferring protein (perhaps as 'spines'?) from the one component to the other. P-protein persists in mature sieve elements (a few dispersed strands are visible in Plate 7, and larger quantities around sieve plates in Plate 8). × 37 000, insert × 80 000.

Plate 24b Later in the development of sieve elements the cytoplasm and vacuole mix, the nucleus (frequently) disappears, and the endoplasmic reticulum metamorphoses to generate the *sieve element reticulum*. Cisternae of rough endoplasmic reticulum (RER) aggregate in stacks, and in the process the ribosomes are lost from all but the outermost faces of the stack (arrows). *Vicia faba*, root tip, × 72 000.

Plate 24c The stacks of sieve element reticulum are usually associated with the plasma membrane. As the contents of the mictoplasm are dissipated, the reticulum becomes more complex in shape. With the eventual loss of *all* ribosomes, it can be difficult to identify and distinguish the intracisternal spaces (IC) from the mictoplasm. In this example five cisternae are seen adjacent to the plasma membrane (PM), compared with which their tripartite construction is thinner (see circled portions of membrane). Cell wall—CW, lumen of sieve element—L. *Vicia faba*, phloem supply to floral nectary, × 200 000.

The functions, if any, of the sieve element reticulum are not understood. One suggestion that has been put forward is that it might prevent mitochondria or plastids from being swept into the sieve plate pores.

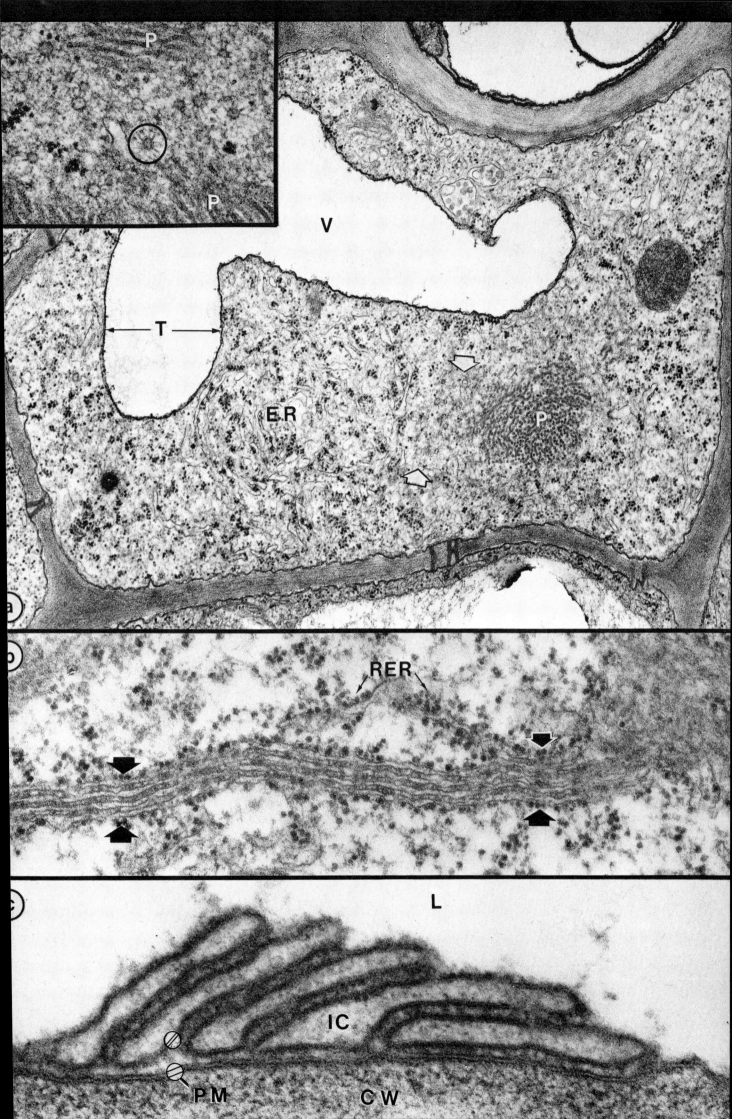

Plate 25

Plates 25-28 present a collection of micrographs illustrating the structure and function of dictyosomes, the units of the Golgi apparatus of the plant cell.

The Membranes of Dictyosomes

Plate 25a The membranes of this freeze-etched dictyosome were revealed by a fracture which passed along an expanse of membrane (upper right), through a collection of vesicles, then steeply down in a succession of narrow steps through 6-8 cisternae, and finally along a more level plane where it exposed face views of a cisterna (centre) and numerous vesicles lying nearby in the cytoplasm.

Comparisons with ultra-thin sections from the same material (cells of the green alga *Micrasterias*) allow identification of the small vesicles at the upper right as transitional vesicles (TV) at the *forming face* of the dictyosome. The first dictyosome cisterna in the stack is pierced by a cleft (arrow), and had the cell not been killed, it is probable that later-formed cisternae would also have been split. In other words, this cleft may be the first sign of a division process which would have resulted in the production of two dictyosomes. The larger vesicles (V) at the bottom and lower left of the micrograph represent the membrane-bound products formed at the *maturing face* of the dictyosome. Other small vesicles lie at the margins of the successive cisternae.

It is difficult, by inspection of the fracture face, to distinguish surfaces that might be external faces of membrane, from surfaces that might represent the internal architecture of a membrane that has been split along its hydrophobic interior.

The extensive surface labelled M probably represents the interior of the membrane of a mature cisterna. Smaller, but similar, views can be seen amongst the ledges above. Other micrographs have shown that, moving from the forming to the maturing face of the dictyosome, the number of particles per unit area of membrane increases. There are about 7000 particles per square μm in the mature cisternae (as at M). The majority of the particles is restricted to the central area of the cisterna, and a marked decline in their frequency is seen towards the periphery, where the membrane surface begins to vesiculate. The vesicles at the maturing face (V) have relatively few particles, a condition that could arise either by removal of the particulate component(s) of the membrane, or by 'dilution' of the particles by extension of the non-particulate areas. $\times$ 82 000.

(We thank Drs. L. A. Staehelin and O. Kiermayer and Cambridge University Press for permission to reproduce this picture from *Journal of Cell Science*, **7**, 787-792, 1970).

Plate 25b and c The forming face of this dictyosome (in another green alga, *Bulbochaete*) is uppermost, where the bifacial endoplasmic reticulum (ER), transitional vesicles (TV), and coalescence of transitional vesicles to form the first cisterna (arrow) can be seen. Vesicles with visible contents are present at the margins of the dictyosome and near the maturing face (V). The upper six cisternae have granular contents, both in the central regions and at their periphery, but in the next three cisternae the central regions are different. There the membranes are clearly tripartite (see (c), a higher magnification view of the same dictyosome), and the intra-cisternal compartment is very thin, thus restricting the contents to the margins, which is where vesicles are formed. In the most mature cisternae the membranes have moved apart again, and they appear to be empty. Presumably the former contents have been packaged into vesicles. The central membranes may break down into fragments (F).

Magnifications: b, $\times$ 70 000; c, $\times$ 155 000, (micrographs provided by T. W. Fraser).

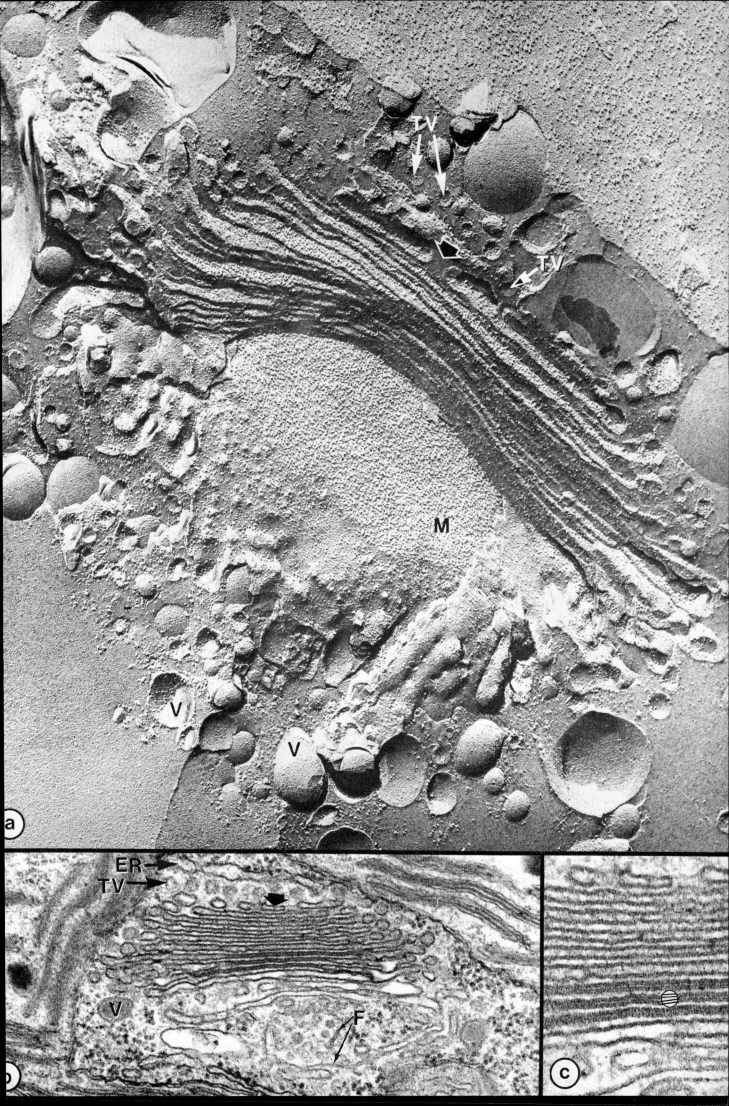

Plate 26

Production of Scales in the Golgi Apparatus

These micrographs, all of members of a genus of planktonic flagellates in the Haptophyceae, called *Chrysochromulina*, illustrate a number of central principles relevant to the functioning of the Golgi apparatus. (a) – (f) depict the end product and (g) – (h) the specificity and mode of its formation.

Plate 26a *Chrysochromulina* produces walls composed of scales. An empty discarded scale case of *C. pringsheimii* is seen here in a shadow-cast whole mount. Two types of scale are visible—the long 'pins' at the ends of the case, and the smaller type, of which a few have broken away from the case (arrow).

Plate 26b - d Plate 26b is an ultra-thin section of *C. chiton*. The scale case in incomplete, nevertheless two types of scale are easily distinguished. Those lying on the outside (O) have a flat base and a curved rim. A shadow-cast example is shown in (d), with the inner face (that closest to the plasma membrane) exposed to view, revealing an intricate pattern of fibrils. The inner scales (I), lying between the outer scales and the cell itself, are slightly curved, with a small recurved rim. In (c) they are viewed from the inside (upper specimen) and also from the outside (lower specimen). The pattern is visible on the inner face of the inner scales, but is overlain with amorphous material on the outer face. The shadow-cast outer face shows the recurved rim, formed from circumferential microfibrils. (b) also shows cell contents: nucleus, N; chloroplast, C, with stalked pyrenoid, P; mitochondria, M; and the single dictyosome, D, with the maturing face at the top.

Plate 26e, f These two sections (along with (a) and (b)) illustrate genetic specificity in the shape of the scales. (e) is a genetically deviant form of *C. chiton*, and differs from that in (b) only in the shape of the outer scales (O), which have the same type of base, but lack the curved rim. The inner scales (I) are like those of 'normal' *C. chiton*. (f) shows outer (O) and inner (I) scales of *C. camella*, in which the outer scales are shaped like cups with four rings of perforations on the sides. Numerous other distinctive species-specific forms could be illustrated.

Plate 26g, h Dictyosomes from the 'normal' form of *C. chiton* (scales as in (b) (c) and (d)), and from the deviant form (scales as in (e)) are shown in (g) and (h) respectively. Both micrographs show: (i) *the dorsiventrality of the dictyosomes*: endoplasmic reticulum (ER), transitional vesicles (between open arrows), and the forming faces are at the *lower* part of the picture. (ii) *formation of scales in dictyosome cisternae*: the first visible signs of scales are indicated by the arrows. Recognizable outer (O) and inner (I) scales are present, each type within its own cisterna. The example of a 'normal' outer scale in (g) shows that the shape of the cisternal membrane matches the shape of the scale (arrowheads). Cisternae at the maturing face, close to the plasma membrane (PM), contain scales of mature appearance. Liberated scales are shown in (h). (iii) *dorsiventrality of cisternae*: in every instance where the plane of section is suitable, it can be seen that the scales in the successive cisternae are oriented in the same way, that is, with their outer faces towards the maturing face of the dictyosome. (iv) *expression of genetic specificity in the Golgi apparatus*: the character which distinguishes the two forms of *C. chiton*, i.e. the shape of the scales, is seen to be developed within the dictyosome cisternae.

The micrographs thus illustrate formation of highly distinctive species-specific structures by the Golgi apparatus. The cisternae are shown to possess individual synthetic capabilities, and, like the dictyosome as a whole, to be dorsiventral.

(a) × 1750; (b) × 10 000; (c) – (h) all × 30 000.

(We thank Professor I. Manton for providing all of these micrographs. (d) and (g) are previously unpublished: (f) is reproduced by permission from *Arch. Mikrobiologie*, **68**, 116, 1969; (a) from *Jour. Marine Biol. Assoc. U.K.*, **42**, 391, 1962; (b), (c) and (e) from *Jour. Cell Sci.*, **2**, 411, 1967; (h) from *Jour. Cell Sci.*, **2**, 265, 1967).

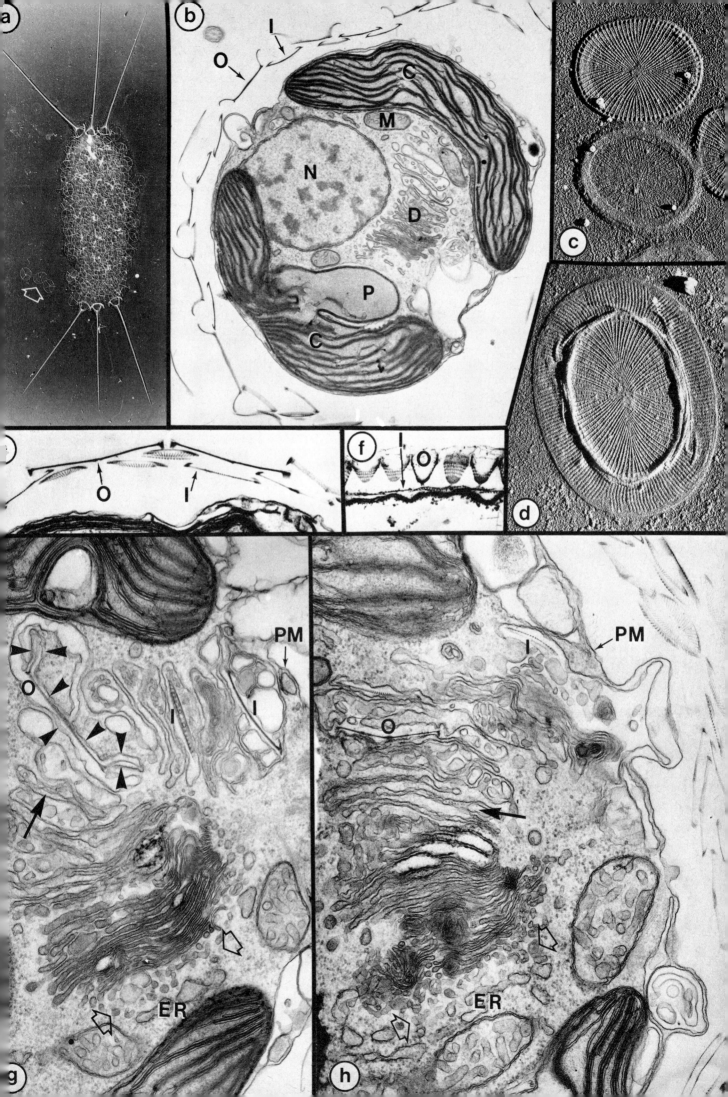

Plate 27

Relationships between Dictyosomes, Endoplasmic Reticulum, and Nuclear Envelope

Spatial relationships between the above components of the cell are especially obvious in certain algal cells.

Plate 27a The filamentous green alga *Bulbochaete* possesses 'hair cells' (chaetae) containing a conspicuous Golgi apparatus consisting of numerous discrete dictyosomes (D 1-3), between which ramify cisternae of rough endoplasmic reticulum (ER). The dictyosomes are oriented with respect only to the subjacent endoplasmic reticulum, thus in D-1 the forming face is lowermost, in D-2 uppermost, and in D-3 at the left hand side. In all three the reticulum is bifacial, lacking ribosomes in the zones (between arrows at D-2 for example) nearest to the forming faces. Transitional vesicles (TV) are present, but so are very numerous other vesicles. The larger type (V), more or less spherical, with granulo-fibrillar contents, are formed at the margins of cisternae at the maturing face. As shown in the insert at top right, their bounding membrane is tripartite (arrow), and of the same dimensions as the plasma membrane. The insert at lower right is part of a section adjacent to that used for the main picture, and it shows dictyosome D-2. Whereas D-2 (main picture) merely shows transitional vesicles (bearing a fuzzy coat on their membrane), D-2 (insert) shows what can best be interpreted as a stage of formation of a coated transitional vesicle (upper arrow) and a stage of coalescence of another such vesicle (lower arrow) with the first dictyosome cisterna.

The endoplasmic reticulum is again bifacial where it lies alongside (between arrows) the vacuole (Va). Hair cells in *Bulbochaete* have no chloroplasts (in fact no plastids of any kind), yet they produce abundant Golgi vesicles, and clearly require raw materials. It may be that the latter enter the hair cell *via* plasmodesmata (PD) from photosynthetic cells on the other side of the cell wall (CW). The vacuole could be a storage reservoir for raw materials, and the closely juxtaposed tonoplast (T) and bifacial endoplasmic reticulum a device whereby entry of nutrients into the reticulum is facilitated. Once in the reticulum, nutrients could pass to the bifacial regions subjacent to dictyosomes. (Micrographs courtesy of T. W. Fraser. Main picture and lower insert × 52 000; upper insert × 120 000.)

Plate 27b In many algae, as here in *Tribonema* at × 60 000, the dictyosomes lie alongside the nucleus, and transitional vesicles (TV) are seen between the nuclear envelope (NE) and the forming face. The examples in the micrograph are used to illustrate presumed stages in formation of a transitional vesicle from the outer membrane of the nuclear envelope, as in the sequence 1-7. The region of the nuclear envelope that produces vesicles has a thick (up to 100 nm) layer of grey-staining fibrillar materials on its outer surface (e.g. star). This is of unknown significance but since it is not seen elsewhere on the nuclear envelope it may be part of an apparatus for vesicle production. For example a system of contractile microfilaments may serve to distort the membrane surface and pinch off the vesicles. (Micrograph provided by Dr. G. F. Leedale, and reproduced by permission, from *Brit. Phyc. Jour.*, **4**, 159-180, 1969).

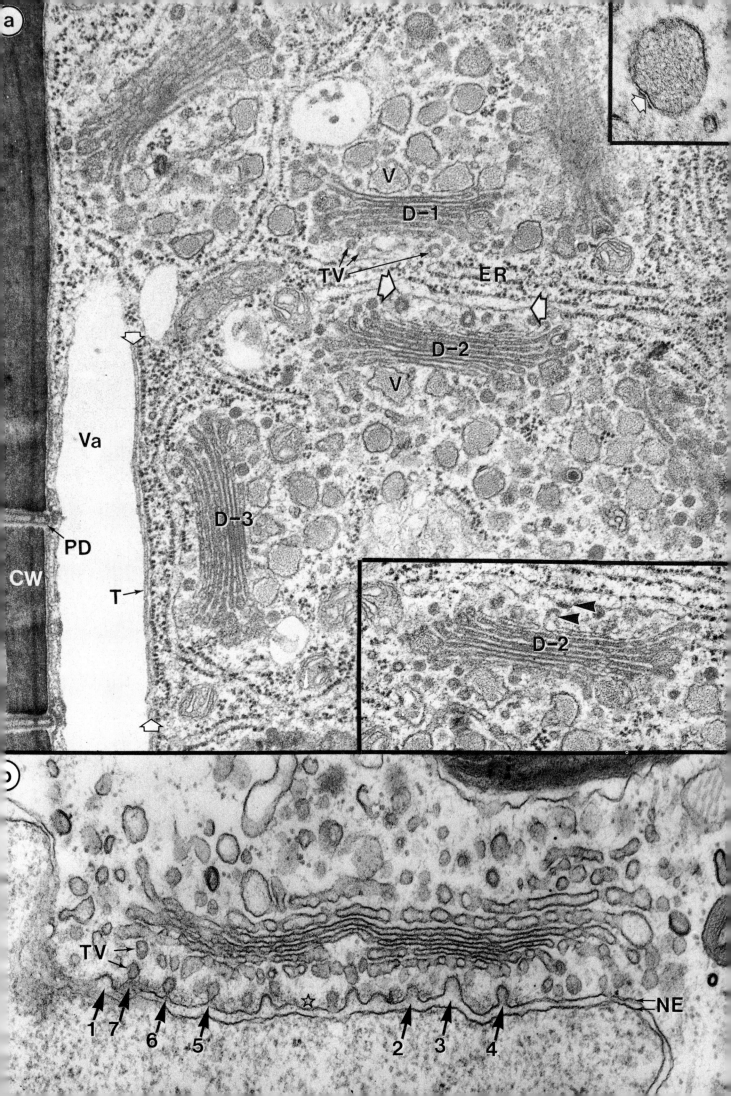

Plate 28

The Golgi Apparatus and Mucilage Secretion by Root Cap Cells

The Golgi apparatus of cells in the root cap has been extensively studied. It mediates the production of a polysaccharide mucilage that is exported by reverse pinocytosis through the plasma membrane and extruded to the surface of the root tip. Two grasses—timothy (*Phleum*) and corn (*Zea*)—are used in this plate because they illustrate different types of vesicle formation.

Plate 28a The periphery of this root cap cell of *Phleum pratense* shows part of the nucleus (N) and the thick, mucilage-laden cell wall (CW). The cytoplasm contains numerous dictyosomes (D), in which the cisternae contain fibrillar material, visible even close to the forming face (F). The cisternae remain flat, with no vesicles at their periphery, and increase in thickness (see numbered sequence in the dictyosome at the right), eventually rounding off at the maturing face to form single, large vesicles, apparently without fragmentation into small vesicles (compare *Zea*, below). Similar large vesicles (V) appear close to the plasma membrane (PM), and the plasma membrane bulges over packages of fibrillar material (e.g. black arrows) that closely resemble the contents of the vesicles. It is reasonable to conclude that within this single micrograph we are seeing many stages in the manufacture, packaging, movement, and delivery of the fibrallar material to its final destination in a more dispersed form (probably as a result of hydration) in the wall (asterisk).

Other noteworthy features include: (1) No obvious relationship between endoplasmic reticulum and the forming face of the dictyosomes, and no identifiable transitional vesicles (compare Plate 27a and b). (2) The forming dictyosome cisternae are composed of branched tubules (T), visible in a dictyosome that lies with its forming face in the plane of the section. The insert shows the appearance of the same forming face in the adjacent section, with extensions of the tubular system. (3) Other micrographs have shown coated vesicles near dictyosomes, and two are seen here at the plasma membrane (circled), either being formed at it or else having fused with it. × 30 000.

Plate 28b The ultra-thin section shown here (the periphery of a root cap cell of *Zea mays*) was treated by a procedure involving oxidation with periodic acid to produce aldehyde groups in polysaccharides; coupling of the reagent thiocarbohydrazide to the aldehydes, and of silver proteinate to the thiocarbohydrazide. The end result is that polysaccharides are rendered electron-dense and so can be located. They occur, as expected, in the cell wall (CW), and conspicuously in dictyosome cisternae, thus confirming the suggestion that the latter are sites of polysaccharide accumulation. The insert shows that the polysaccharide does not have the same appearance in the cisternae and nearby vesicles as in the wall (bottom right hand corner of insert), thus indicating that a maturation, or perhaps more simply a hydration, process takes place between the two locations.

Vesicle formation at the maturing faces of *Zea* dictyosomes is not the same as in *Phleum* (above). The dictyosome under the insert shows stained polysaccharide in the successive cisternae and then, at the maturing face, its migration to form *peripheral* vesicles (solid arrows), leaving a fragment of membrane (open arrow) derived from the central region of the cisterna. As in *Phleum*, there are no clear associations between endoplasmic reticulum and dictyosomes. × 15 000, insert × 36 000.

(Micrograph (b) and insert provided by Dr. M. Rougier; (b) reproduced by permission from *Journal de Microscopie*, **10**, 67-82, 1971).

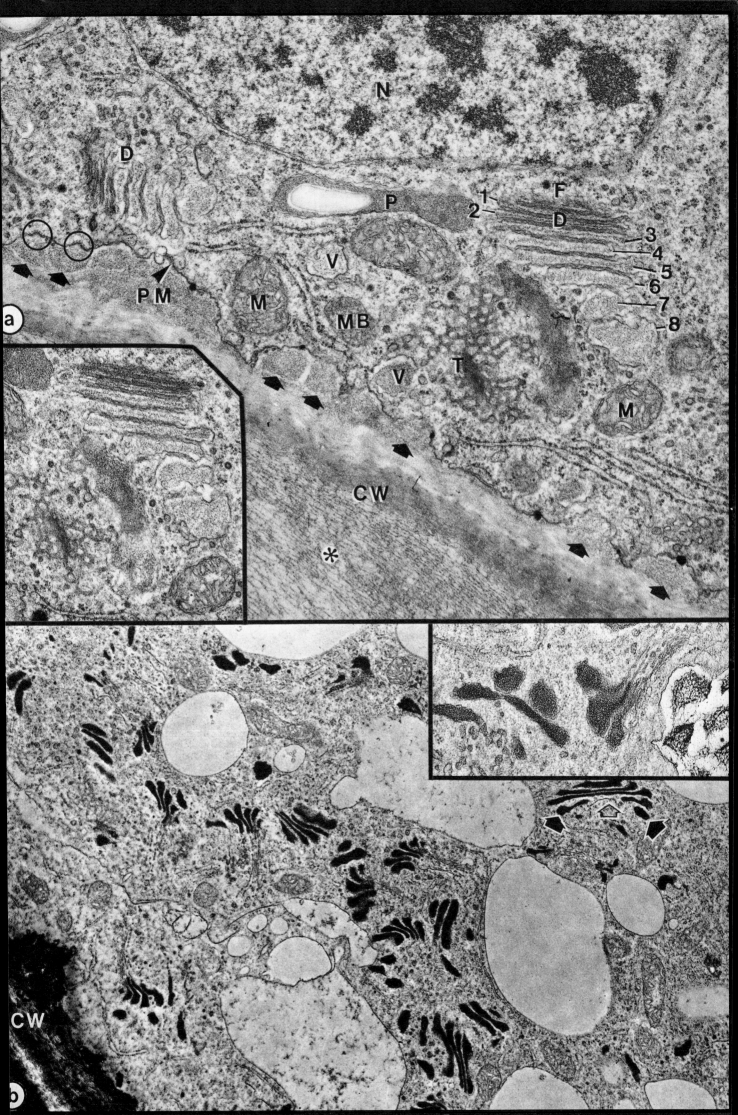

Plate 29

Mitochondria (1)

Mitochondria are constituents of all plant cells, and their number and conformation in a cell are related to the respiratory activity of that particular cell. The micrographs on this page illustrate some of the more commonly encountered forms, as seen in cells having a range of respiratory activities.

Plate 29a This group of mitochondria is lying in the cytoplasm of a parenchyma cell within the floral nectary of broad bean (*Vicia faba*). Although it is difficult to ascribe a precise function to this particular cell, the tissue as a whole secretes nectar, a solution containing mainly sucrose, in an energy-requiring process involving active transport across the plasma membrane. Most of the mitochondrial profiles seen here are circular, suggesting that the mitochondria approximate to spheres 0.75 – 1.5 μm in diameter. Individual mitochondria are seen to be surrounded by an outer (O) and an inner (I) membrane, the latter being infolded (arrowheads) to form the cristae (C). In the matrix of many of the mitochondria there are electron-transparent areas, the nucleoids (N), which contain fine DNA-fibrils (F). Mitochondrial ribosomes (R) lie in the more densely stained regions of the matrix, but are inconspicuous. Mitochondrial granules (G) are more obvious, and probably consist of calcium phosphate. One of the mitochondria (asterisk) is linked to another (star) by a continuous outer membrane (O). This could represent a stage of mitochondrial division, but the same appearance could also be obtained in certain sections of a Y-shaped mitochondrion.

Other features are the plastid (P), and the dictyosome (D) (seen in face view with its associated vesicles). There are numerous dictyosome vesicles, some quite large (DV), and containing a fibrillar material. Many of the free cytoplasmic ribosomes are organized in poly-ribosomal helices (PR) (see also Plate 21c). × 28 000.

Plate 29b This transfer cell (see also Plate 13) is located alongside two xylem elements (X) in the cotyledonary node of a lettuce seedling (*Latuca sativa*). The cytoplasm adjacent to the wall-membrane apparatus contains many mitochondria with densely packed cristae (C), which nearly completely obscure the nucleoids (N); otherwise they are similar to the mitochondria in (a) above. The high rate of respiration indicated by the number and conformation of these mitochondria may be connected with consumption of energy in the pumping of solutes across the plasma membrane of the transfer cell. × 24 000.

Plate 29c Many cells are less active (in terms of respiration) than either nectary or transfer cells, and their mitochondria are correspondingly simpler, as illustrated here using part of a young endodermal cell of an *Azolla* root. Although the mitochondria are similar in size and shape to those in (a) they have a much reduced system of cristae (C). The outer (O) and inner (I) membranes are quite distinct, but neither shows the tripartite construction as, for example, found in the plasma membrane. The outer membrane (and the cytoplasm in general) is densely stained, probably as a result of liberation of phenolic compounds during fixation. The mitochondrial ribosomes (R) are smaller than their cytoplasmic counterparts. As in the other micrographs on this plate, the number of profiles of cristae that are seen far exceeds the number of visible connections between the cristae and the inner mitochondrial envelope (e.g. open arrows). The inference is that the connections are small, and that the cristae expand from small 'necks' into the central compartment of the mitochondrion. × 106 000.

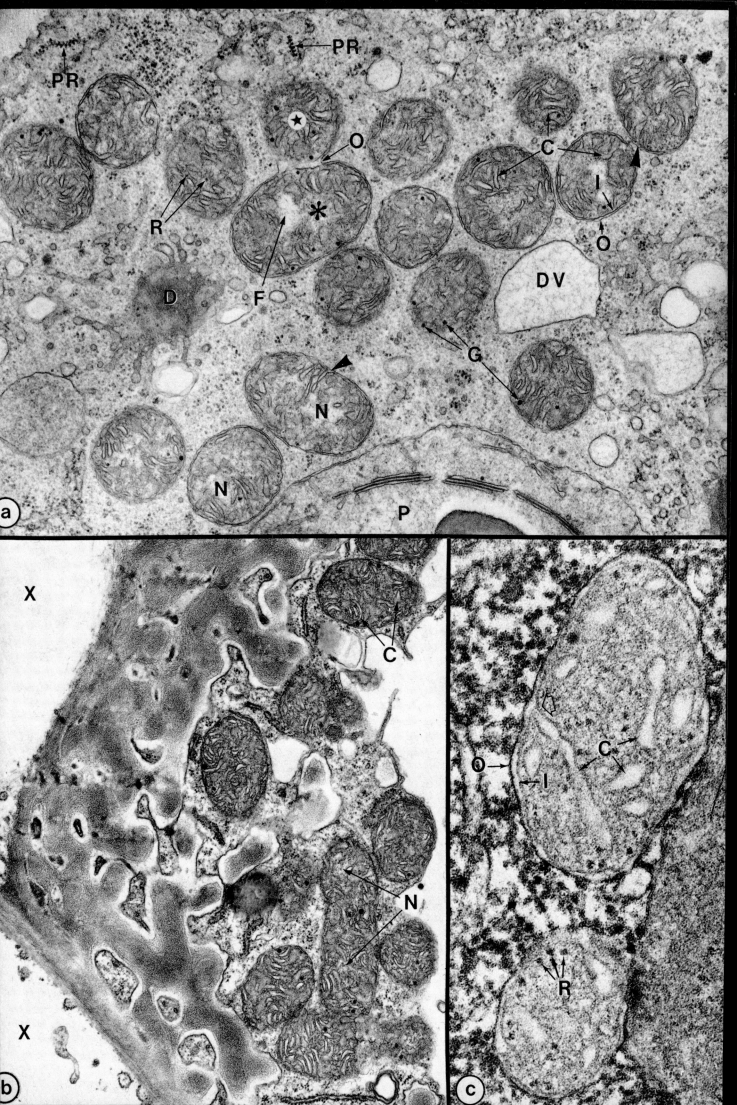

Plate 30

Mitochondria (2)

Although the examples presented in Plate 29 are typical of mitochondria in many plant cells, other configurations can also be found. Some are displayed here.

Plate 30a These unusual mitochondria were found in some root tip cells of white lupin (*Lupinus albus*). Interpretation of the three-dimensional configuration is clearly impossible from a single micrograph. One mitochondrial profile (M-1) forms an irregular but continuous ring, with another (M-2) lying quite close to it. M-1 may indeed be ring-shaped, but it is more likely that both it and M-2 are parts of a single plate that undulates up and down through the plane of the section. Support for the latter proposal comes from other micrographs such as the insert, which demonstrates that arms of the mitochondrion are connected (asterisk). If all parts of M-1 and M-2 are connected in this manner, the plate so formed must be at least 15 μm × 4 μm. The insert also shows the numerous small cristae (C) and mitochondrial ribosomes (MR) (smaller than cytoplasmic ribosomes (CR)) lying in the mitochondrial matrix. Magnifications: × 16 000, insert × 36 000.

Plate 30b and c The micrograph shown in (b) (× 20 000) is one of a sequence of 43 that together encompassed the whole of a young cell of the unicellular green alga *Chlorella fusca* var. *vacuolata*. The complete sequence was used in building the three-dimensional reconstruction of the mitochondrion and chloroplast shown in (c). The model is viewed as from the bottom of (b) looking upwards; (b) corresponds to the level in the model marked by the arrows X and Y. Components are labelled as follows: nucleus (N – dotted line in (c), nucleolus seen in (b)); vacuoles (V – containing electron-dense polyphosphate bodies); chloroplast (C – containing a pyrenoid (P) which is shown exposed in (c) by cutting away the layer of chloroplast that covers it); microbody (MB – lying in the cytoplasm, but in close proximity to the pyrenoid, a regular feature of these cells); centriole pair (CP – not included in (b)).

The main feature displayed in the reconstruction is that the mitochondrion (M), seen as quite separate profiles (1-8) in (b), is a 3-dimensional continuum of loops and branches – a mitochondrial reticulum. Some branches lie close to the cell surface (2 and 3 in (b) = A and B in (c)). Another branch (7 in (b), C in (c)) penetrates a cytoplasmic channel that runs through the chloroplast, emerging at D. Very few cells or organisms have been reconstructed in three dimensions, and it is therefore not known how commonly the mitochondrion is in the form of a reticulum.

The three layered (dark-light-dark) structure (W) in (b) is the remains of the wall of the parent cell. In this (but not all) species of *Chlorella*, the layer of the parent cell wall that survives contains a substance closely resembling sporopollenin, the resistant material of pollen grain walls (Plates 11, 12).

Plate 30d One bizarre mitochondrial conformation that is commonly found is a ring-shaped profile that probably results when a cup-shaped mitochondrion is sectioned across the 'cup'. The two layers of the mitochondrial envelope are visible at both inner *and* outer faces of the ring (circles). In this example, from a developing cotyledon of *Phaseolus multiflorus*, the crista is unusual in that it is very extensive, perhaps even forming a continuous cisterna within the mitochondrion. × 60 000.

Plate 30e In contrast to the mitochondria of normal vegetative cells these four examples, from a developing pollen grain of *Avena fatua*, are very small (none is more than 0.5 μm in diameter) and relatively undifferentiated. Each contains a few cristae (C) and mitochondrial ribosomes (R) within the inner (I) and outer (O) membranes. Both mitochondria and plastids can become very rudimentary during development of pollen. × 55 000.

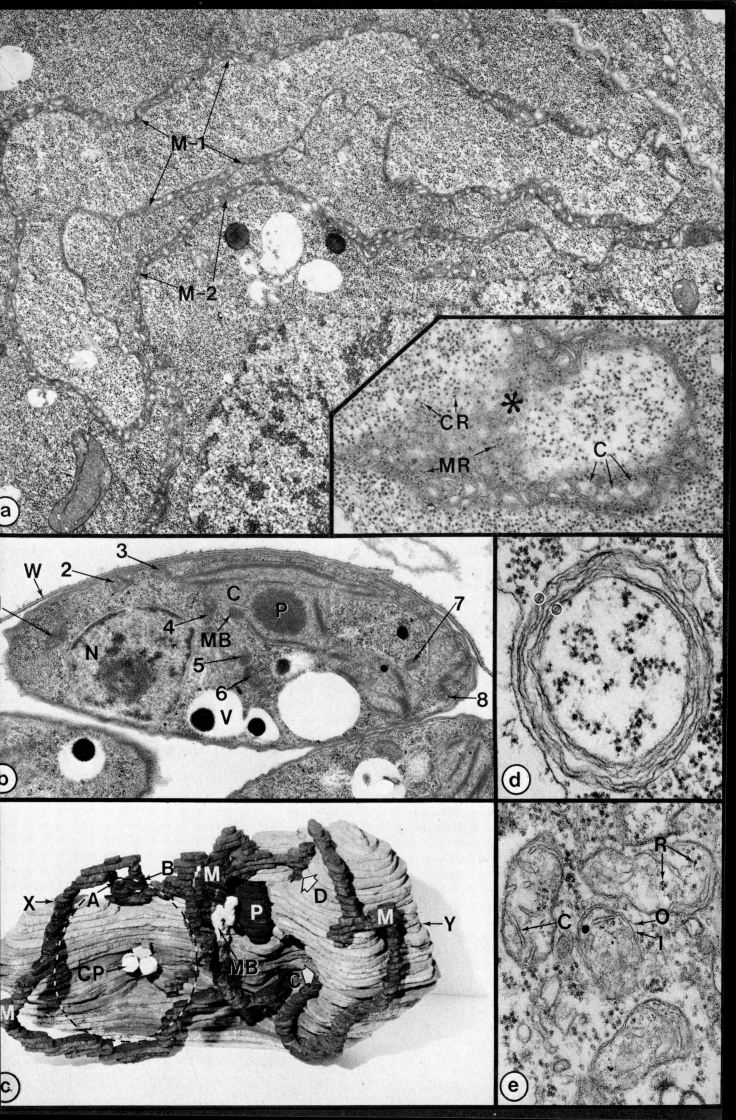

Plate 31

Plates 31–40 are concerned with the structure, development, and inter-relationships of plastids.

Plastids I: Proplastids and their Development to Etioplasts and Chloroplasts

Plate 31a Plastids with rudimentary internal structures, but with the capacity to develop into more complex types, are described as proplastids. They are illustrated here (P) in a cell at the outer surface of a stem apex, and in part of an underlying cell (bottom left of micrograph). They are similar in both cells, despite the fact that maturation and differentiation of the two will produce very different plastids—rudimentary types in the mature epidermis, and well developed chloroplasts in the underlying tissue.

A nucleus (N), with nucleolus, a large vacuole (V) with flocculent contents, several smaller vacuoles, and the thin cuticle (C) are also seen. Oat (*Avena sativa*) stem apex, × 5300.

Plate 31b The proplastids of meristematic cells at the stem apex (31a) are similar to those in root tip cells. Root tip proplastids, however, do not normally develop into chloroplasts. The one shown here displays the two concentric membranes of the envelope (visible, e.g., within the circles), with occasional invaginations of the inner membrane (arrows, left hand side). A small starch grain (S) is present in the section, partly ensheathed by internal membranes (thylakoids). The internal membrane system is sparse and not organized. A few particles which may be plastid ribosomes (smaller than cytoplasmic ribosomes) are seen (e.g. within square) and nucleoid areas (large arrows), with fine fibrils, presumably containing DNA, are also present in the stroma. *Vicia faba* root tip, × 77 000.

Plate 31c This micrograph is representative of the appearance of cells which are differentiating in the light from the meristematic condition shown in 31a to become mesophyll cells. The vacuoles (V) are large, and there are intercellular spaces (IS) in the tissue. Much of the chromatin in the nucleus (N) is heterochromatic (dense areas).

The plastids (P) have developed complex internal membrane systems, and by now are young chloroplasts, though grana are not yet obvious. Some starch grains are present in them (S).

Avena sativa, young leaf of illuminated seedling, × 8000.

Plate 31d The same tissue as for 31c was used to obtain this micrograph, but the seedling was grown in darkness so that it became etiolated. Vacuoles (V), intercellular spaces (IS), and nucleus (N) are much as in the light-grown seedling, but the plastids have developed into young etioplasts (E). The main feature of etioplasts is the semi-crystalline lattice of membranous tubes known as the prolamellar body (open arrows). The example numbered 1 is of the type illustrated further in Plate 37h, and number 2 is of the type shown in Plate 36e and g. Number 3 is probably of the same type as Plate 37g. If the plant were to be illuminated, etioplasts would pass through developmental stages of the type shown in Plate 38, and become chloroplasts.

Avena sativa, young leaf of etiolated seedling, × 8000. Reproduced by permission from *Biochemistry of Chloroplasts*, II, 655–676 (Ed. T. W. Goodwin, Academic Press, 1967).

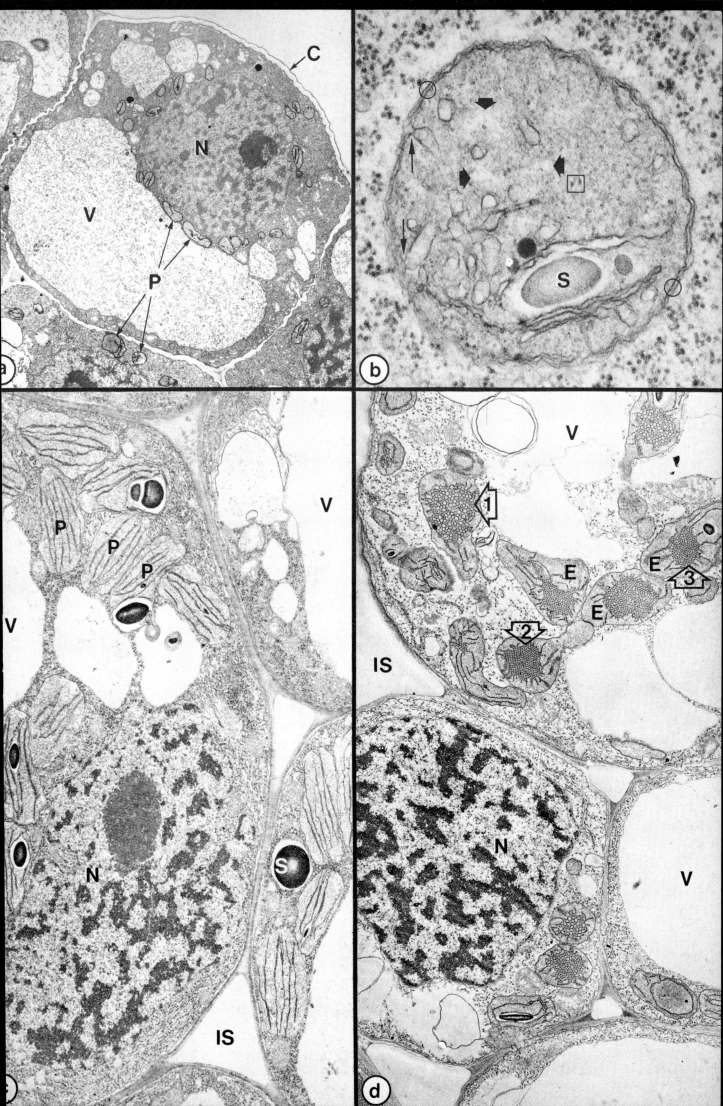

Plate 32

Plates 32-35 illustrate chloroplasts.

Plastids II: Chloroplasts (1)

Plate 32a A transverse section of a typical leaf is viewed here by light microscopy. The cells of the upper and lower epidermis (lowermost and topmost cell layers, respectively) do not contain well developed chloroplasts. These are visible, however, in cells of both the spongy mesophyll (S) and the palisade layer of columnar cells (P). Two small veins are included, and in the one on the right, it can be seen that chloroplasts are not obvious in the bundle sheath cells (stars) (compare Plate 34a). *Hypochaeris radicata*, × 165.

Plate 32b This is a high magnification light micrograph of the same material as in (a). One palisade cell is shown. The section (1-2 μm thick) has passed through the peripheral layer of chloroplasts in the cell. Numerous grana (the dark grains) are resolved within each chloroplast. The hemispherical shape of the chloroplasts can be seen from the two types of profile present, the 'side views' corresponding to the plane shown in (c), the 'top views' to that in (d). × 1100.

Plate 32c and d The electron microscope reveals finer details of chloroplast structure: the two membranes of the chloroplast envelope (E); 'side' (S) and 'top' (T) views of grana; densely staining ribosomes in the stroma (circled) (note in (c) that they are smaller than cytoplasmic ribosomes); starch grains (G), which for some unknown reason have become stained very differently; the system of fret membranes (F) that interconnect the grana (better seen in (c) than in (d), where most of them are present in oblique or face view); plastoglobuli (P). (c) *Avena ventricosa*, × 33 000; reproduced by permission from *Canadian J. Genetics and Cytology*, **12**, 21-27. (d) *Zea mays* mesophyll chloroplast, × 18 500.

Plate 32e and f Further details of *Zea mays* grana are illustrated here in 'top view' (e) and 'side view' (f). The grana contain very many more layers (discs) than those in (c), and the stroma (S) is very much subdivided by the frets (F) passing between the grana. A few ribosomes (circled) can be seen in the stroma. The fret channels are somewhat swollen in (f), making their presence and their course easier to distinguish. In (e) the frets are in oblique view, and they appear merely as grey shadows, some of which are marked. The granum at top right appears to have been sectioned in the plane of the discs. The striations across the other grana indicates how oblique the section is in each case—each striation representing an oblique slice through one disc. Numerous frets (white arrowheads) can be detected entering the granum at top right, and since the section is only thick enough to accommodate about 3 granum discs, this means that *each* disc must develop *many* connections to frets. (e) × 55 000, (f) × 75 000.

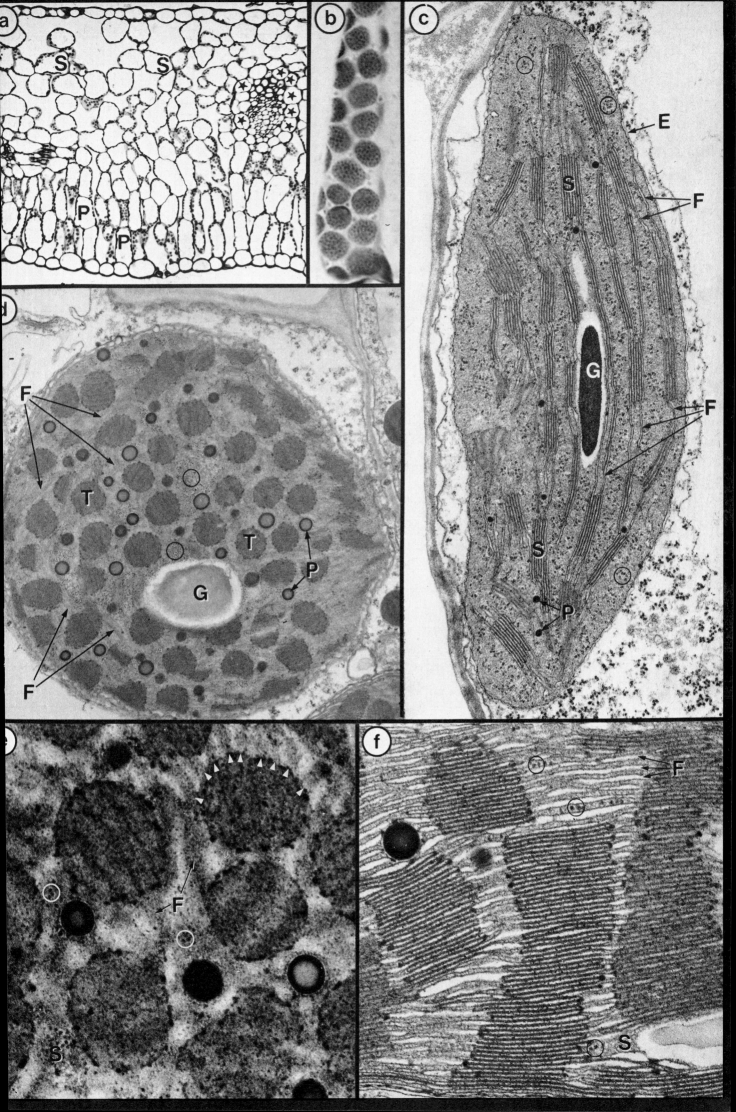

Plate 33

Plastids III: Chloroplasts (2): Details of Chloroplast Membranes

Plate 33a This ultra-thin section of part of a lupin (*Lupinus* sp.) leaf chloroplast shows several details of both the chloroplast envelope (E) and the internal membrane system. The inner of the two membranes of the envelope is invaginated in places. No subunits can be seen in the membranes, but the 'A-space' between granum discs is visible where the plane of the section is precisely at right angles to the plane of the discs (white arrowheads). Where the section has included the points at which fret connections enter granum discs, continuity between the fret channels and the disc loculi is shown (open arrows). × 140 000.

Plate 33b Grana with very numerous discs are examined here by freeze etching. The fracture plane has passed down two grana (G, lower left and upper right) approximately at right angles to the plane of the discs. A third granum is in the centre of the micrograph, and it lay obliquely to the fracture, so that surface views of membranes were exposed. The clear white bands are 'steps' where the fracture has descended from one membrane to another. The intervening particle-studded areas represent the interior of the disc membranes. One face carries large particles (the B-face), the other larger numbers of smaller particles (the C-face). It can be seen that on passing from the discs to the system of frets that interconnects the 3 grana, the large B-face particles become less frequent (arrows from labelled B-face). The small C-face particles do not diminish in numbers from disc to fret (arrows from labelled C-face). *Alocasia*, × 67 000.

Plate 33c In this high magnification view of a freeze etched preparation, about two thirds of the area of a granum disc membrane is seen in the bottom part of the picture, adorned with the scattered large particles typical of the B-face. Numerous frets radiate away from the circumference of the disc, and the fracture has exposed examples of both their B- and their C-faces (B and C respectively). Several fret B-faces are continuous with the B-face of the disc (arrows), and as in (b), the frequency of the larger particles is seen to be much lower in the frets. By contrast, the fret C-faces have very numerous small particles. Judging by the number of large particles present, the area at the upper left of the picture may be the edge of another granum, in which small areas of many successive B- and C-faces are exposed. *Lomandra longifolia*, × 105 000.

(We thank Dr. D. Goodchild for providing (b) and (c).)

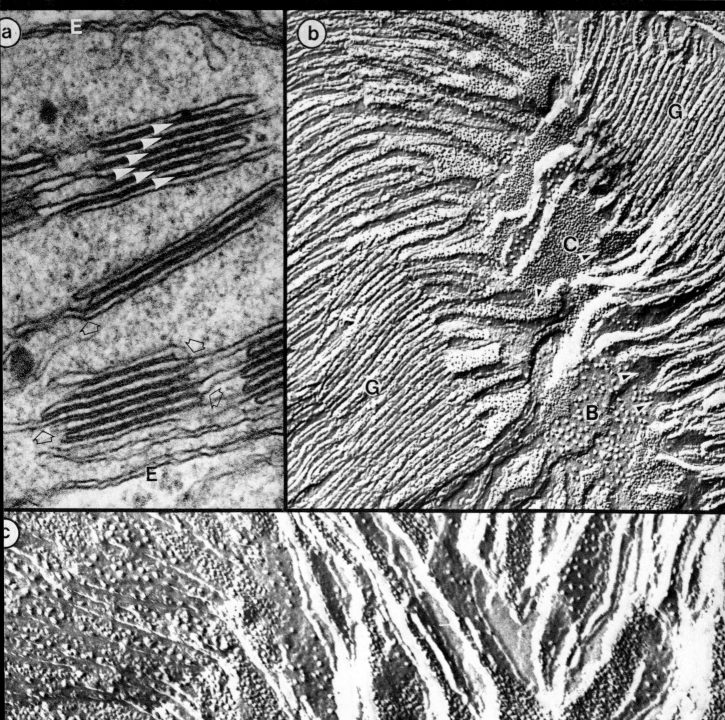

Plate 34

Plastids IV: Chloroplasts (3): Dimorphic Chloroplasts in the C-4 Plant, Zea mays

Plate 34a The two types of chloroplast found in C-4 plants are distinguishable even in the light microscope, as here in a section of a maize leaf. The xylem (X) and phloem (P) of each vein are surrounded by a ring of bundle sheath cells—5 in this example. The bundle sheath is in turn surrounded by mesophyll cells. Grana show up darkly stained in the mesophyll chloroplasts (e.g. at large arrowheads). Starch grains are unstained (e.g. small arrows), and are abundant in the agranal bundle sheath chloroplasts. × 1750.

Plate 34b Bundle sheath (BS, lower right) and mesophyll (M, upper left) chloroplasts are compared in this electron micrograph. The former contain starch grains (G) in the stroma between the simple, agranal, internal membranes. The latter possess grana and frets (seen in more detail in Plate 32). Both types have chloroplast ribosomes (smaller than cytoplasmic ribosomes), plastoglobuli, and the usual double membrane envelope. The leaf was still growing when it was fixed, and the chloroplasts had not completed their development, as indicated by the presence of a small region of prolamellar body lattice (solid arrow), perhaps a relic of the membrane growth that had taken place in darkness during the night before the early morning harvesting of the material.

The cell wall passing diagonally across the micrograph contains the layer of suberized material that surrounds each bundle sheath cell (open arrows). Plasmodesmata (P, parts of two groups included) interconnect the bundle sheath and mesophyll cell protoplasts. The tonoplast (T) of each cell can be seen, separated from the vacuolar face of the chloroplasts by only a very thin layer of cytoplasm. × 27 000.

Plate 34c Parts of bundle sheath (BS, on left) and mesophyll (M, on right) chloroplasts are shown here at higher magnification, along with the intervening wall, with its suberized lamella (open arrows, here obliquely sectioned) and a group of plasmodesmata (P). Both of the chloroplasts possess a feature not shown in the previous figure—invagination of the inner of the two envelope membranes to form a 'peripheral reticulum' in the chloroplast. The invaginations (large solid arrowheads) lead into somewhat dilated sacs, pierced by many perforations through which the stroma penetrates (small arrows). This structure is not restricted to C-4 chloroplasts.

The bundle sheath chloroplast does not have a completely 'unstacked' membrane system. Two and three disc rudimentary grana are seen (stars), survivals from early development when grana were larger and more abundant. × 64 000.

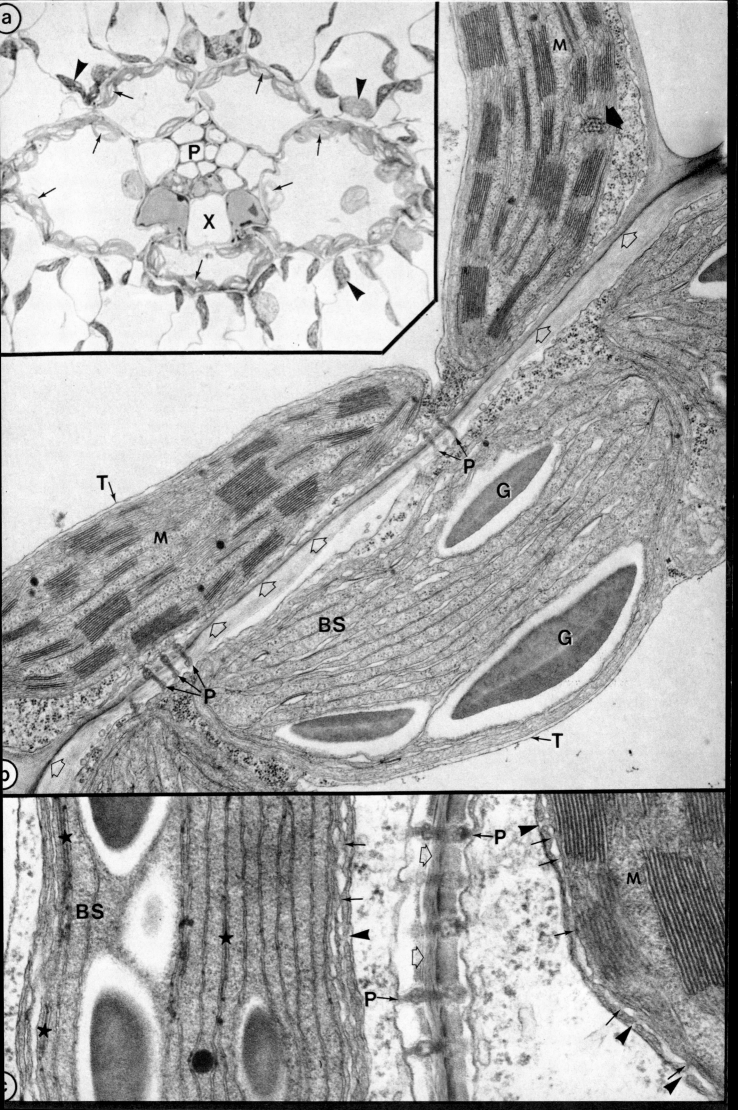

Plate 35

Plastids V: Chloroplasts (4): Components of the Stroma

Plate 35a Several components of the stroma are displayed in this micrograph of a greening oat leaf (see also Plate 38c, d).

Ribosomes: More densely stained than the surrounding material, the plastid ribosomes may be scattered or in chains (rectangle) or clusters (between stars). Some lie in contact with, and probably are bound to, the thylakoids (arrowheads). They are smaller than cytoplasmic ribosomes, seen at lower left outside the plastid envelope (E).

Nucleoid: The nucleoid (N) sectioned here contains fine fibrils, 2-3 nm in thickness (small arrows). These are probably the histone-free DNA of the plastid. Other fibrillar material is present, as are granules (double arrows) which could be developing plastid ribosomes.

Proteinaceous ground substance: The general background in the plastid is finely particulate (except in the nucleoid, which presumably contains some material that excludes the rest of the stroma). The particles cannot be identified, but many must consist of the CO_2-fixing enzyme, ribulose 1,5-diphosphate carboxylase, some molecules of which are shown negatively stained (pale against a dark background) at high magnification in the insert (one example ringed). The enzyme is a very large protein, with a molecular weight of half a million, each molecule being a particle of side about 10 nm.

Crystals and spherulites (ordered aggregates) of proteinaceous material are quite commonly found in the stroma, particularly when the plastids have been subjected to physiological stress. The example shown here (S) occurs in *Avena* plastids under natural conditions. It consists of bundles of fibrils, which in turn are built up of small units, perhaps molecules of ribulose diphosphate carboxylase. × 94 000, insert × 430 000.

Plate 35b The iron-containing protein, ferritin, occurs in most types of plastid, occasionally in conspicuous masses, as here in a lupin leaf chloroplast. Each electron dense point represents the several hundred iron atoms present in the centre of each protein molecule. The protein shell around the iron core is not visible. Some parts of the mass of molecules show a measure of symmetry. When ferritin crystallizes in the plastid the arrangement of molecules is as shown in the insert.

Several dark, spherical plastoglobuli are present at the periphery of the ferritin aggregate. × 120 000.

Plate 35c-g Chloroplasts of most algae contain one or more pyrenoids, which are easily seen, even with the light microscope, and in the green algae are rendered especially conspicuous by the development of ensheathing grains of starch. In the electron microscope the pyrenoid is seen as a dense granular mass, occasionally with a crystalline lattice, and in many species traversed by modified thylakoids. The example shown here is simple, with no thylakoids. The sequence of pictures illustrates changes that occur during the life of a *Chlorella* cell. In very young cells (C) starch (solid stars) appears as a thin plate around the pyrenoid (P) and accumulates while the cell photosynthesizes and grows (d, e). Starch grains also appear in other parts of the chloroplast stroma (open stars in d, e and f) but the pyrenoid starch is especially labile. It is eroded *before* the non-pyrenoid starch as the cell begins to divide (f), and the products of the division generally have starch-free pyrenoids (g).

(c) and (d) illustrate (as does Plate 30b, c) the proximity of the microbody (M) to the pyrenoid in this alga. All × 33 000, except (f), at × 18 000. (Micrographs provided by Dr. A. W. Atkinson, Jr.)

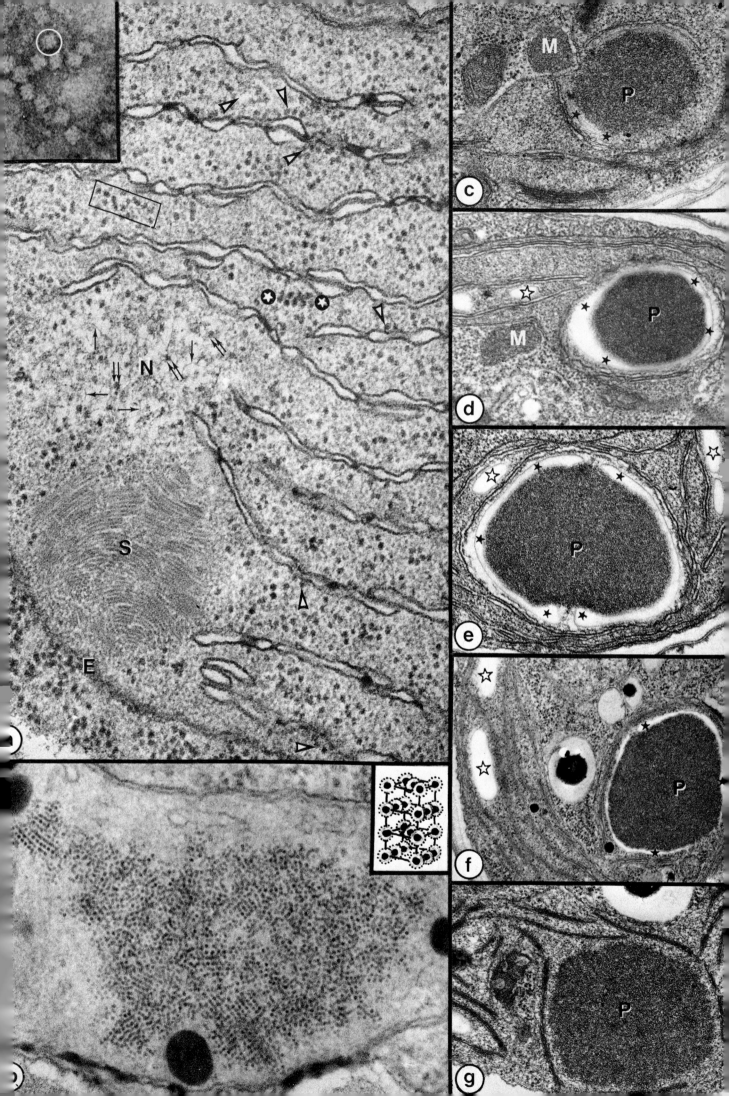

Plate 36

Plastids VI: Etioplasts and Prolamellar Bodies (1)

Etioplasts develop in leaves on plants that are growing in darkness. Their main distinguishing feature is the system of membranes known as the prolamellar body. The micrographs and models on this and the next plate illustrate some of the diversity of form of prolamellar bodies. All of the examples are from leaves of oat seedlings (*Avena sativa*).

Plate 36a These two etioplasts display the two membranes of the plastid envelope (black arrow), and numerous plastid ribosomes, which are somewhat smaller than their cytoplasmic counterparts. The prolamellar bodies (PLB) can be seen to be semi-crystalline lattices, but in these examples the plane of section is not such as to display regular lattice planes to advantage. Subsequent micrographs have been selected to do this. The lattice at the bottom centre region of the left hand prolamellar body (above the white arrow) is obviously different from the neighbouring lattice (see also Plate 37h). × 47 000.

Plate 36b, c, d One, uncommon, form of prolamellar body lattice is illustrated here by a micrograph, a 3-dimensional drawing, and a model. The lattice is composed of tubes, branched and interconnected in three axes at right angles to one another. The unit of construction is shown enclosed within the square on the micrograph. The etioplast stroma penetrates the prolamellar body *between* the tubes, but in (b) little structure is seen in the stroma component of the prolamellar body because uranyl acetate stain (which adds contrast to ribosomes) was not used. Models of the type shown in (d) will be used to illustrate the other types of prolamellar body (below): comparison with (c) emphasizes that such models represent only the orientation of the tubes, and *not* their diameter and smoothly confluent surface contours. (b) × 78 000.

Plate 36e, f Most prolamellar body membrane lattices consist of tetrahedrally branched tubular units. The arrangement of the tubes in the lattice shown here is analogous to the crystal structure of wurtzite (a mineral form of zinc sulphide). The view of the model in (f) (left hand side) matches an area such as that marked on the micrograph. The other views of the model are side views—looking at an edge (centre) or at a face (right hand side) of the hexagonal 'crystal'. (e) × 70 000.

Plate 36g, h The micrograph and photographs of models depict an alternative lattice form, analogous to diamond crystals or the zincblende form of zinc sulphide. The overall crystal shape is bi-pyramidal. The views of the model were taken looking at an edge (left hand side), at a face (centre) and at a vertex (right hand side). The major difference between this lattice and the previous is that successive planes of hexagonal 'rings' are out of register. The centre picture of the model illustrates the overlapping hexagons of the lattice (compare (f) (left hand)). The section (g) is somewhat tilted relative to (h) (centre), but individual hexagons are seen in horizontal bands, each band being part of one plane of the lattice. The insert (of the area above the arrow at lower centre of the main picture) shows how the successive bands are out of register by half a hexagon. This does *not* apply to the lattice shown in (e) and (f).

Both (e) and (g) show the many etioplast ribosomes in the stroma component of the lattice. They also show continuity of lattice tubes and flattened thylakoids projecting outwards from the prolamellar body. (g) × 70 000, insert × 120 000.

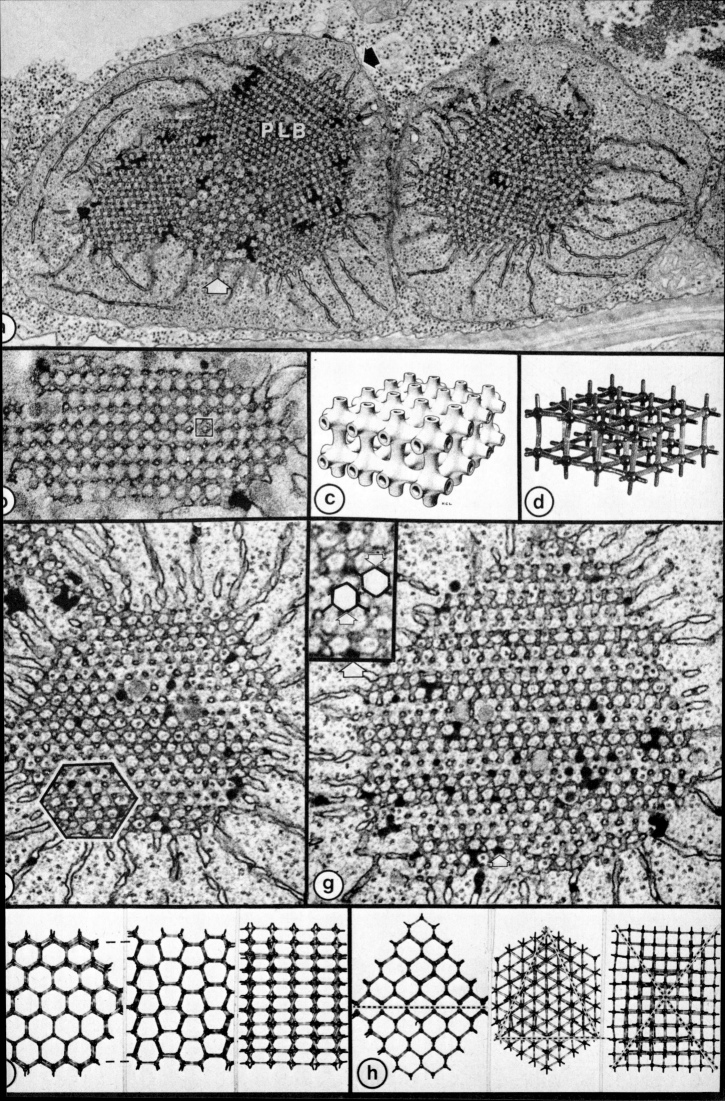

Plastids VII: Prolamellar Bodies (2)

Plate 37a Tetrahedrally branched models of carbon atoms were used (as in Plate 36f, h) to construct this large tetrahedron-shaped constructional unit, which is the basis of the prolamellar bodies in (c), (d) and (g). It consists of 'zincblende' type lattice (as Plate 36g, h) but with a pentagonal dodecahedron at each vertex, and special 5- and 6-membered rings running along each edge. The latter arise when the large tetrahedral units join together in prolamellar bodies.

Plate 37b, c, d, e 'Centric' prolamellar bodies have 20 large tetrahedral units radiating outwards, one from each of the 20 vertices of a central pentagonal dodecahedron (seen in section at the centre of (c) and (d). In three dimensions, an icosahedral shape is generated, as seen in (b), where a model is viewed looking straight at one vertex (the edges of the tetrahedral units on the near side of the model have been drawn in). In median section (to which (c) and (d) *approximate*) parts of 10 of the large tetrahedral units are seen, each one radiating out from the centre. The 10 sectors are marked in the micrographs by the straight white lines. For comparison, a median slice of the model is shown in (e). Ribosomes are seen in the lattice. (c) × 38 000; (d) × 73 000.

Plate 37f, g Here the large tetrahedral units (see (a)) are combined in a more complex fashion—not just radiating out from a central pentagonal dodecahedron, but filling space between many evenly spaced pentagonal dodecahedra. (f) represents a slice (rather thick in relation to the ultra-thin section seen in (g)) through a model of this form of packed tetrahedra. One pentagonal dodecahedron is at the centre, 4 others are marked (arrows) at the periphery, and yet others were above and below the plane of this slice. The triangular sectors at upper left and lower right in the model represent triangular faces of units of the type seen in (a).

In the micrograph (g), lines indicate the boundaries of triangular faces of the type seen in (f). Just as the model could have been extended, this prolamellar body extends to upper left. Also certain areas are missing (e.g. pentagonal dodecahedra at or near positions indicated by asterisks. (g) × 84 000.

Plate 37h, i, j, k There is symmetry in this very complex form of prolamellar body, but the sections are so much thinner than the basic units of construction that the details of the lattice are difficult to discern. An area approximating to (j) is marked on the micrograph. Much of the lattice consists of rows of pentagonal dodecahedra, interconnected at 60° to one another to form a network containing gaps which create the hexagonal pattern seen in (j). Rows inclined at 60° to one another can be seen if the micrograph is held up and viewed at a shallow angle along the three axes parallel to the sides of the white hexagon. Side views of the lattice (viewed edge-on in (i), and face-on in (k)) show how the successive strata of pentagonal dodecahedra (between brackets) are joined by other types of ring.

This type of prolamellar body can occur in isolation, or it may be joined to one of the other types (as in Plate 36a). (h) × 70 000.

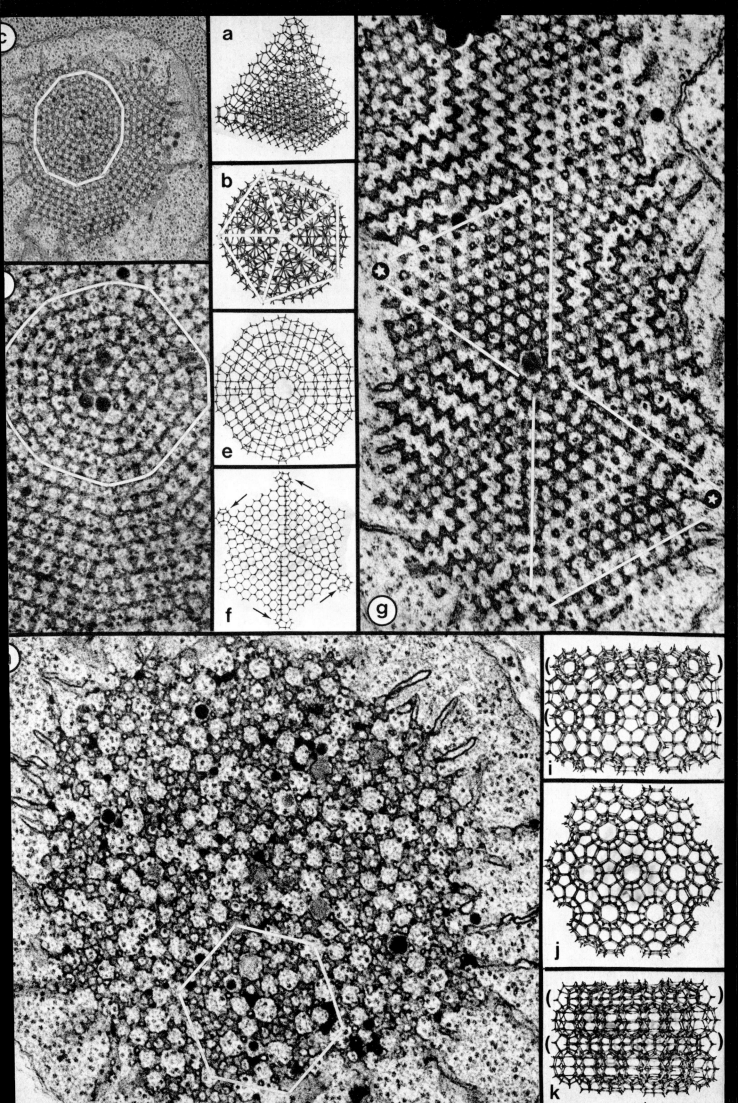

Plate 38

Plastids VIII: The Greening Process: From Etioplast to Chloroplast

Plate 31 illustrates the development of proplastids to chloroplasts in plants grown in the light, and their very different development to etioplasts in plants that are kept in darkness. Subsequent plates deal with chloroplasts (32–35) and etioplasts (36, 37) in more detail. The present plate shows three stages in the conversion of etioplasts to chloroplasts, as seen when a dark-grown plant is illuminated. As in Plate 31, the micrographs are of portions of oat (*Avena sativa*) leaf. Each stage is illustrated by means of sections which show thylakoids in profile view (a, c, e) and in face view (b, d, f).

Plate 38a, b Illumination very rapidly converts the protochlorophyll found in the prolamellar body of etioplasts to chlorophyll. Much slower structural changes are also set in train. The prolamellar body loses its crystallinity and the membrane in it metamorphoses into spaced out flattened sacs called primary thylakoids. This pair of micrographs represents a stage reached 2 hours after the dark-grown plants were brought into the light. The prolamellar body remnant (PR) has not completely dispersed. Many of the perforations that were delimited by tubular membranes in the prolamellar body still survive as small pores through the primary thylakoids (small arrows in the profile (a) and face (b) views of the thylakoids). The plastid ribosomes are mostly in clusters and chains suggestive of polyribosomes (e.g. open arrows). Invaginations of the inner plastid envelope membrane (stars) and a nucleoid area (N) are present in (a).

Plate 38c, d The amount of protochlorophyll that is converted by light to chlorophyll is very small compared with the amount of chlorophyll found in a mature chloroplast. Net synthesis of chlorophyll begins in the greening plastids after a lag period, which in the material shown here lasts 2–3 hours, and during which no new chlorophyll, and probably no new membrane, is produced. The beginning of the period of rapid synthesis is marked by the appearance of portions of membrane overlapping (large arrowheads) the primary thylakoids. This, the first stage of granum formation, is shown in both profile (c) and face (d) view (and at higher magnification in Plate 35a). One of the overlapping discs is sectioned through its fret connection (opposed arrowheads). Most of the perforations seen in (a) and (b) have by now disappeared from the primary thylakoids, though a large prolamellar body remnant still survives (right hand side of (c)). The membranes of the primary thylakoids are still continuous with those of the prolamellar body (e.g. at small arrows).

Plate 38e, f After 10 hours in the light the leaves are obviously green, but still not as green as they would be if they contained mature chloroplasts. Chlorophyll and membrane synthesis has progressed. The small overlaps seen in (c) and (d) have extended to become full sized granum discs (see (f)), and new disc-shaped layers have been added, producing small but clearly recognizable grana, interconnected by frets derived from the primary thylakoids (see (e)). A mass of plastoglobuli marks the remnant of the prolamellar body in each picture.

(c), (d), (e) and (f) all show that chains of ribosomes lie on the surface of the growing membrane (beside asterisks in all four micrographs), very like the polyribosomes found on the rough endoplasmic reticulum. These 'rough thylakoids' may be synthesizing protein molecules which, when complete, pass directly into the growing membrane surface.

Magnification × 36 000 in all except (c), at × 64 000.

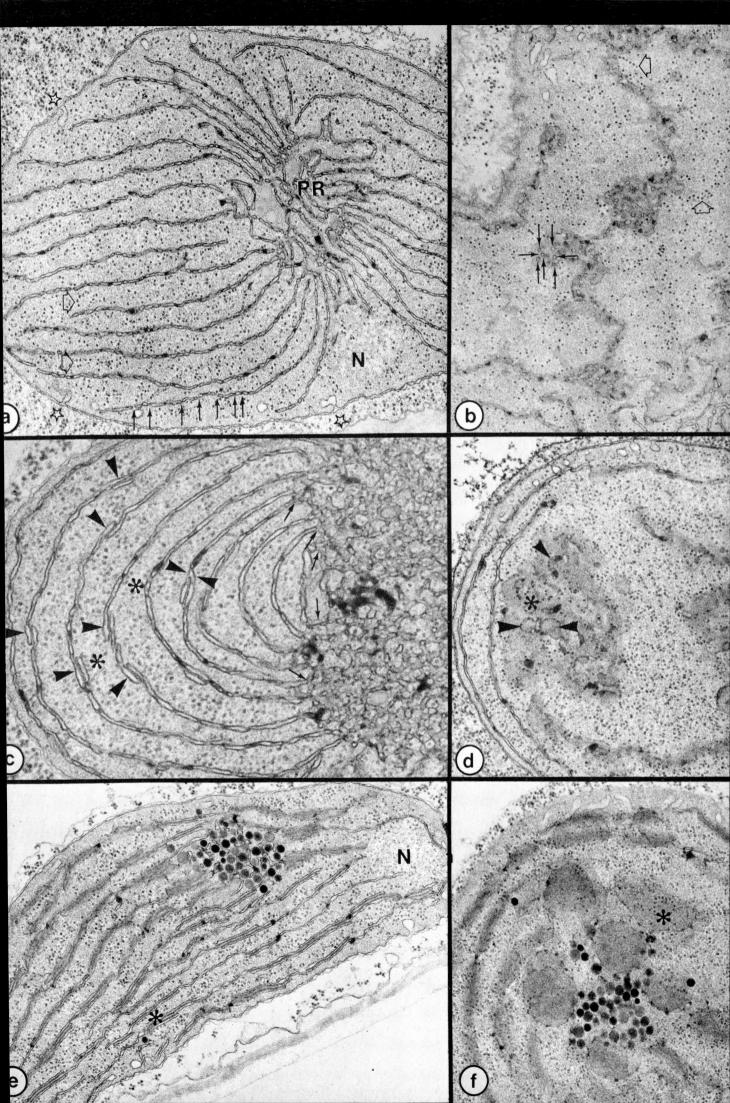

Plate 39

Plastids IX: Amyloplasts

Plate 39a, b These two scanning electron micrographs show cells in a piece of potato tuber that was prepared by conventional fixation and then dried by the 'critical point' method, in which distortion is minimized. A freshly-cut surface was exposed before taking the pictures. The cells are highly vacuolate in life, indeed some parts of those in (a) appear to be nearly empty. The most conspicuous of the cytoplasmic components is the population of starch grains. The grains are ovoid when large, and nearly spherical when small.

The cell at the bottom of (a) is seen at higher magnification in (b). Amyloplast envelope membranes are not resolved; they lie closely appressed to the starch grains. The largest grains are about 30 μm $\times$ 50 μm, and the smaller spheres about 3 μm in diameter. Numerous strands of cytoplasm, some with remains of more particulate cell components, form a network stretching out to a peripheral thin layer of cytoplasm at the cell wall. The gaps between the strands presumably represent vacuoles which in life were traversed by the strands of cytoplasm. (a) $\times$ 280; (b) $\times$ 680.

Plate 39c These three amyloplasts from a peripheral cell of a soybean (*Glycine*) root cap are very much smaller than the reserve amyloplasts in (a) and (b) and Plate 20a. Many starch grains (S) are present in each. General features of plastids that can be seen include the double membrane envelope (black arrow) and nucleoid areas (white arrow). The internal membrane system is not well developed, but stacked thylakoids occur occasionally (star). Thylakoids sometimes lie closely appressed to starch grains (open arrows), but it is not known whether this reflects membrane-activity in starch metabolism, or whether the starch grains merely became pressed against the membranes as they grew. $\times$ 23 000.

Plate 39d The starch in sieve element plastids is of an unusual type, containing a high proportion of branched chains of glucose. Whereas ordinary amylase would digest potato starch grains, pretreatment with a special 'de-branching' enzyme is necessary to break down sieve element starch. The grains are unusually electron-dense, and display a granular composition approaching that of glycogen (which is also a branched polymer of glucose). The double envelope can be seen around the upper grain. *Coleus* petiole sieve element, $\times$ 15 000.

Plate 39e These amyloplasts grouped round the nucleus (N) in a young root cap cell of *Cosmea* each contain numerous starch grains (usually round in shape), and in addition accumulations of material that is extremely dense to electrons after processing for electron microscopy. The accumulations lie in distended intra-thylakoid compartments: in other words they are not in the stroma, where electron-dense plastoglobuli are found. In other material it has been found that this type of accumulation can be digested away from the section by treatment with the lipid- and protein-digesting enzymes lipase and pronase. It may therefore contain a lipoprotein. Phenolic material could also be present. Equally dense deposits are seen in the vacuoles (asterisks) and there is evidence (again from other material), that plastids can extrude phenol-containing droplets to the cytoplasm and vacuoles. Other features of the micrograph include mitochondria (M) with cristae and small dense granules, and lipid droplets (L). $\times$ 26 000.

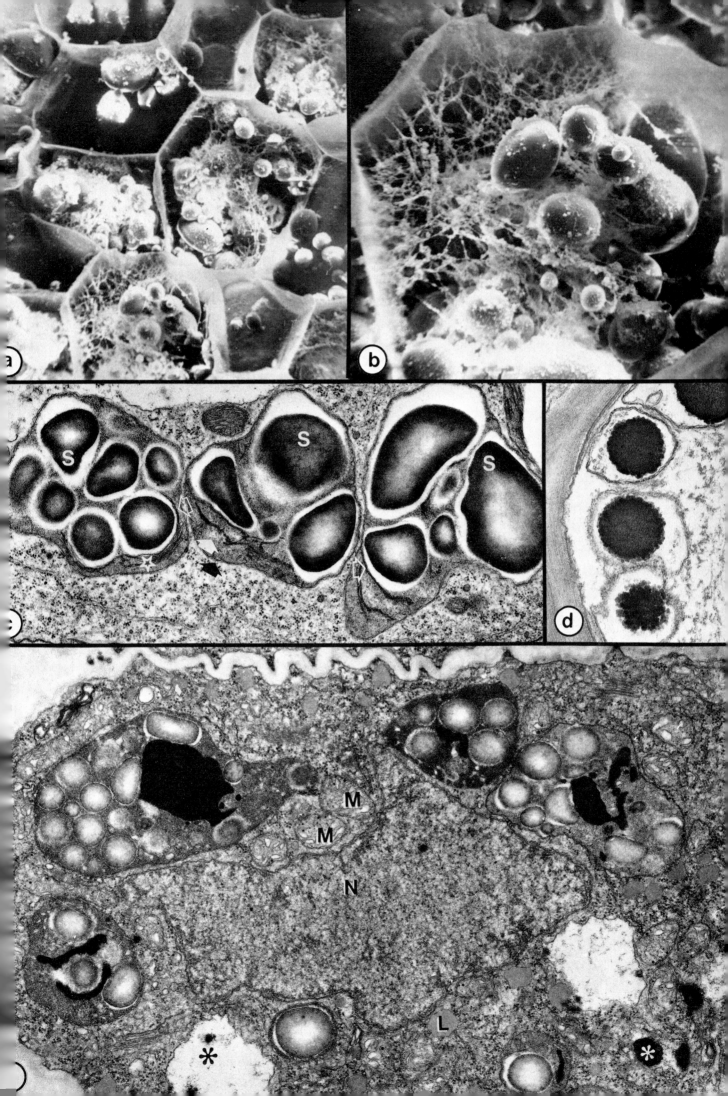

Plate 40

Plastids X: Chromoplasts

Plate 40a Chromoplasts are illustrated here in part of a cell from the orange rim of the corolla tube of a *Narcissus poeticus* flower. The upper left portion of the micrograph is occupied by part of the nucleus, which is seen to contain much dense heterochromatin, and to have a high ratio of surface area to volume, being penetrated by long cytoplasmic channels (asterisks) which are lined by pore bearing (arrows) nuclear envelope. The cytoplasm is vacuolate (V), and contains mitochondria (M), cisternae of rough endoplasmic reticulum, numerous free ribosomes, and lipid droplets (L).

The chromoplasts dominate the cytoplasm. Their outlines, and especially their internal membranes, are convoluted. The clear zones within them (stars) represent what in life were crystals of beta-carotene. They are now electron transparent, some at least of the carotene having been extracted during dehydration of the specimen after the shape of the crystals was preserved by fixation. Numerous membranes undulate through the electron transparent areas. Many of the chromoplasts contain electron dense globules—plastoglobuli, which may, like the crystals, contain chromoplast pigment, but probably also contain (as in chloroplasts) plastid quinones. × 15 000.

Plate 40b The chromoplast shown here (from a tomato fruit) is in a relatively early developmental stage. Its juvenility is shown by the presence of many small grana: later in development these disappear, leaving electron dense plastoglobuli and crystals of lycopene (lycopene is a precursor of beta carotene, which also occurs in tomato chromoplasts, but (except in certain varieties) in concentrations that are apparently too low to give extensive crystallization).

The lycopene crystals (stars) have angular outlines, and are surrounded by membrane. They also contain undulating membranes, sometimes aggregated in electron dense stacks (crystal in chromoplast at left hand side). In these respects lycopene crystals resemble beta-carotene crystals (see (a)).

Other features shown in the chromoplasts include a round membrane-bound inclusion (arrow), aggregated electron dense material (open arrow, possibly remnants of a starch grain, or components of plastoglobuli undergoing crystallization). The stroma contains plastid ribosomes, and nucleoid areas (large circle). The double envelope of the chromoplast is also visible (small circles).

Tomato fruit parenchyma cells are very large and vacuolate (V, vacuole; T, tonoplast). A microbody is included in the micrograph (asterisk). × 32 000.

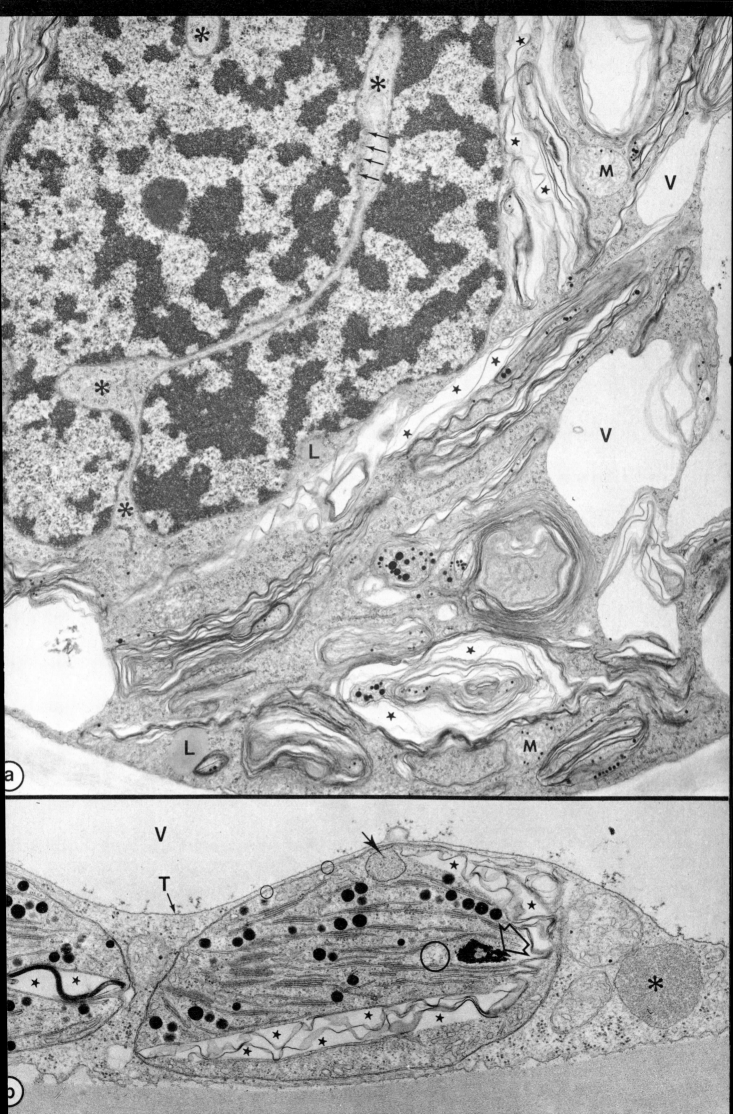

Plate 41

Microbodies

Microbodies are found in various biochemically distinguishable categories, two of which, peroxisomes and glyoxysomes, are illustrated here.

Plate 41a The microbody (MB) in this section of a tobacco leaf is of the type known as a peroxisome, which functions in the enzymatic processing of glycolic acid, produced by chloroplasts, and broken down in peroxisomes yielding carbon dioxide together with other substances which can be retrieved and utilized by the plant. Microbodies (in general) are bounded by a single membrane (arrows). The peroxisome shown here contains a large crystal (CY) surrounded by a diffusely granular matrix. The single membrane is closely appressed to the outer membranes of the two adjacent chloroplasts. The latter contain grana (G), frets (F), and ribosomes (CR) which are smaller than their cytoplasmic counterparts. Note the invagination (I) of the inner membrane of the chloroplast envelope. Also present in the cytoplasm is a mitochondrion (M). The electron-transparent area to the upper right is part of the large vacuole.

The presence of the enzyme catalase in the peroxisomes has been demonstrated. After normal fixation with glutaraldehyde pieces of tissue were incubated with 3.3′—diaminobenzidine and hydrogen peroxide. Visualization and stabilization of the precipitate produced by catalase activity in the presence of these substances was achieved by routine post-fixation with osmium tetroxide, whereupon the sites of enzyme activity became heavily stained. The insert shows a typical peroxisome from tissue treated in this way. Enzyme activity occurs throughout the peroxisome, but is concentrated in the crystal (CY). It is of interest that where the peroxisome membrane abuts on to the neighbouring chloroplasts (arrows), there is a stronger staining reaction than elsewhere. It may be that in these regions of close contact, there is an especially high rate of transport of molecules between chloroplast and peroxisome. × 41 000; insert × 44 000.

(Micrographs courtesy of Drs. S. E. Frederick and E. H. Newcomb, reproduced by permission of the Rockefeller University Press from the *Journal of Cell Biology*, **43**, 343 (1969)).

Plate 41b and c Microbodies in the cells of cotyledons of lipid-storing seeds (e.g. the sunflower illustrated here) provide a striking example of the versatility of microbody activity. During early stages of germination and seedling growth the lipid reserves are broken down with the aid of enzymes present in the microbodies (MB). Some of the enzymes comprise the 'glyoxylate cycle', so this type of microbody has been called the glyoxysome. They lie alongside the lipid droplets (L) in the cotyledon cells. 41(b) represents a stage four days after germination. After all or most of the stored lipid has been consumed the cotyledons enlarge, become green and photosynthetic, and function as leaves. 41(c) represents such a stage, seven days after germination. The microbodies (MB) in (c) are in the same cells as those in (b), but have become associated with the newly developed chloroplasts (C). The glyoxysome type of microbody has been replaced by (or has metamorphosed into) a peroxisome type of microbody. The microbodies in these cells have a somewhat irregular shape, and, as in 41a, the prominent single membrane bounds a dense granular matrix, which in the present examples contains a crystal (CY).

The lipid droplets in (b) possess a surface skin of half membrane (arrows) which is probably formed as a result of the orientation of those lipid molecules that come into contact with the aqueous environment of the cytoplasm. (b) × 31 000; (c) × 34 000.

(Micrographs courtesy of Drs. P. J. Gruber and E. H. Newcomb, reproduced by permission of Springer-Verlag from *Planta*, **93**, 269 (1970)).

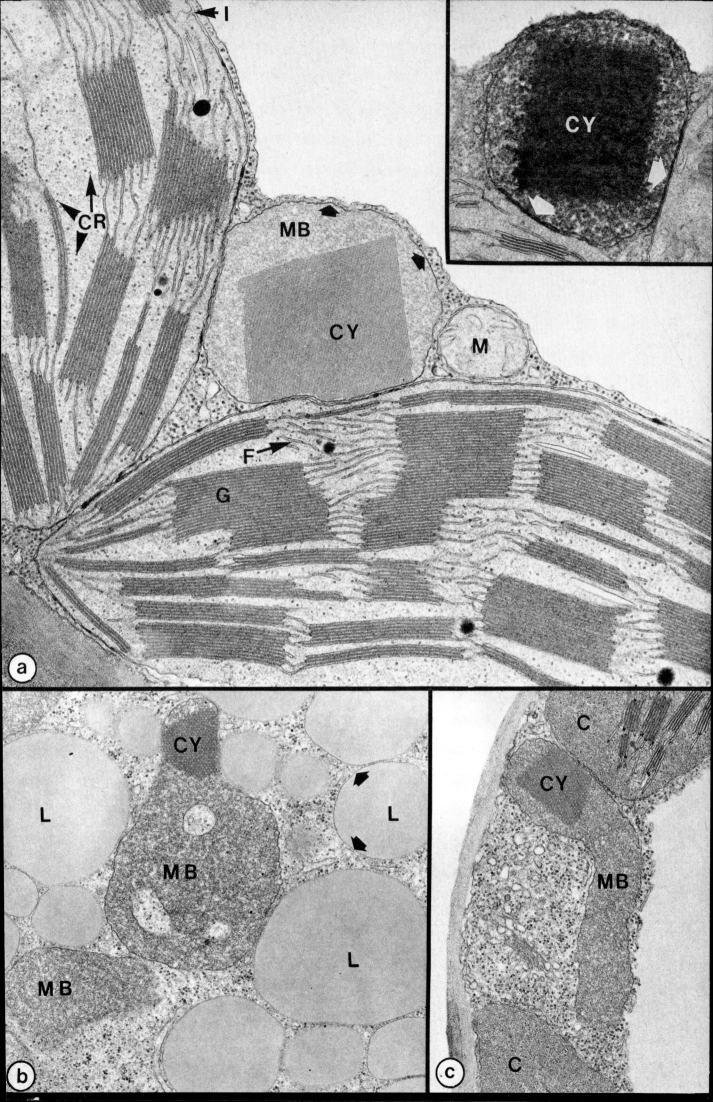

Plate 42

Plates 42 and 43 show aspects of the microtubules of non-dividing cells. Those of cells that are undergoing cell division appear in Plates 44-48.

Cortical Microtubules

Plate 42a One location where higher plant microtubules are found in non-dividing cells is in the cell cortex, just beneath the plasma membrane. Here they may be seen in transverse (T) or longitudinal (L) section. The tubules have a densely staining wall about 7 nm thick, surrounding a clear lumen. Often there is a clear zone around the outer perimeter of the microtubule, from which ribosomes (etc.) are excluded. Cortical microtubules may be linked to one another, and to the plasma membrane, by short 'bridges' (not shown here). *Azolla* root tip, × 54 000.

Plate 42b Sections tangential to the cell surface include the cell wall (CW), plasma membrane (PM) and the immediately underlying cytoplasm (CP). A major feature that is often visible in such planes of section is the correspondence between the orientation of microtubules and that of cellulose microfibrils in the cell wall on the other side of the plasma membrane. In this section from a spinach root tip (*Beta vulgaris*) the wall microfibrils (MF) are clearly visible running horizontally across the lower part of the micrograph (lower arrow). Nearby there is a similarly oriented microtubule (1). Above this there is a band of microtubules (2) running at a slight angle to (1), and again the adjacent wall microfibrils parallel this orientation (arrow at right centre). A further band of microtubules (3) runs at an angle across the main group.

The irregular nature of the cell surface (also shown in (c)) generates very complex images in tangential sections. Numerous dark-light-dark profiles of sectioned membranes are visible (see circled areas, for example), corresponding to sections through the sides of bumps and hollows in the plasma membrane. The irregular electron-transparent areas correspond to similar areas seen in transverse sections of the wall (as in (c)). Many vesicles (V), possibly of dictyosomal origin, lie in the peripheral layer of cytoplasm, amongst the microtubules. × 50 000.

Plate 42c Just before prophase of mitosis it is common for microtubules to aggregate in a bundle lying close to the wall in an equatorial position in the cell. The aggregate is known as the preprophase band, and an example may be seen here between the large arrowheads. The cell wall was very much longer than the small portion illustrated in this micrograph, and the cortical cytoplasm along its length contained no microtubules except for those in the preprophase band. The bumps and hollows of the wall and plasma membrane seen in (b) are also shown here, but in transverse rather than tangential section. Cabbage root tip, × 50 000.

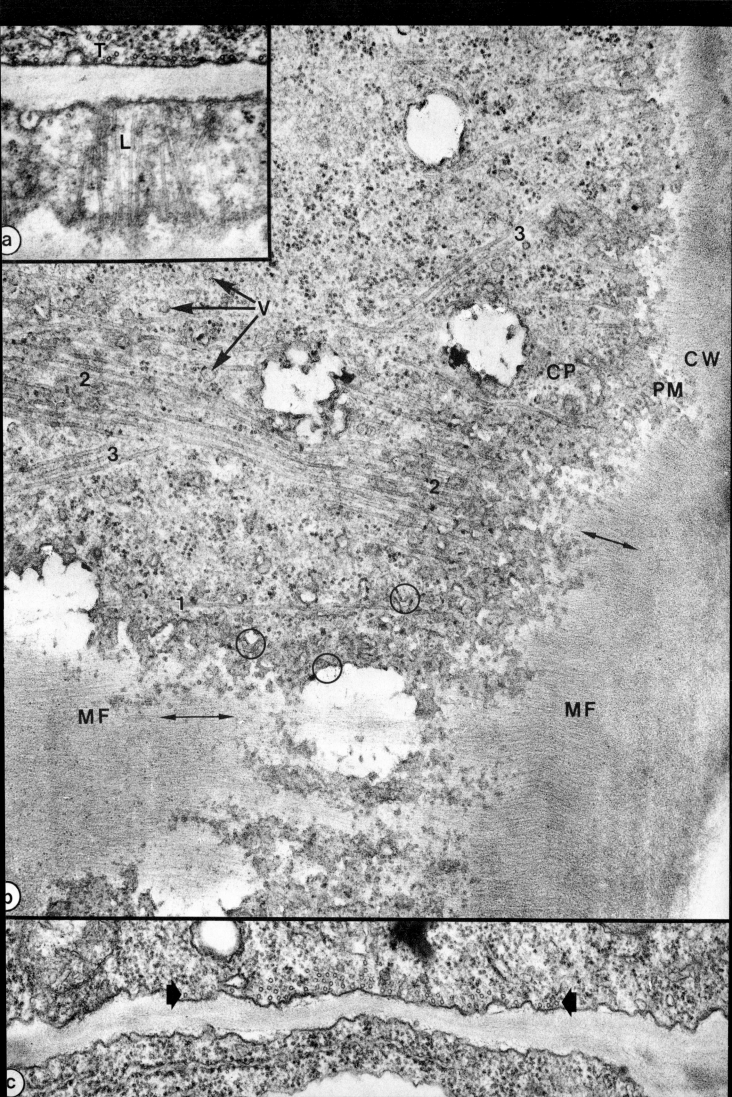

Plate 43

Microtubules and Microfilaments

Plate 43a Certain cells accumulate compounds which can bring about progressive modification of the normal staining patterns after processing for electron microscopy (see also Plate 3b). The appearance of microtubules is especially subject to this phenomenon. A relatively minor effect is seen in the microtubules (MT) to the left of the cell wall in (a). Normally their walls are stained uniformly (as in Plate 42), but here they are somewhat uneven, probably because densely-staining material has become bound to them during fixation. The cell and the microtubules to the right of the cell wall are more severely affected. Dense material has been deposited around *and within* the microtubules, and in interstices of the microtubule wall itself, which appears pale against the dark interior and exterior. *Azolla* root tip, × 132 000.

Plate 43b - e Deposition of electron-dense material around and within the microtubule leads to a form of 'natural' negative staining, revealing details of microtubule substructure that are not normally visible. One microtubule from (a), (circled) is shown enlarged in (b) to × 860 000, and its wall clearly consists of a ring of subunits. It is not possible (in this picture) to count the precise number of subunits in the circumference. The information is, however, present in the image, and can be extracted. The image of each subunit is photographically superimposed upon that of each of its neighbours. This accentuates the regular periodicity around the circumference, while randomly distributed points in the image tend to cancel one another. The subunits can be counted when the picture is rotated just the right amount between photographic exposures to give perfect superimposition. A trial-and-error procedure is used, rotating the picture through a range of angles (e.g. for 4 subunits, 90°; for 12 subunits, 30°). Superimposed images formed by this method are shown in (c), (d) and (e). The three pictures represent tests aimed at discovering whether the tubule in (b) has 12, 13 or 14 subunits around its circumference. Only one, (d), gives an image that is free from distortion and blurring. In the others the angle of rotation did not correspond to the angle subtended by the subunits, so that no true image reinforcement occurred. The conclusion is that the wall of the microtubule contains 13 circumferentially placed subunits.

Plate 43f In non-dividing cells microtubules may be found elsewhere than in the cell cortex, associated in a skeletal function with cell components other than the plasma membrane. In this example it may be that they anchor the nucleus (N) in position in a specific region of the cell. The section was tangential to the nucleus, thus enabling the pores (NP) and the polyribosomes (P) of the nuclear envelope to be seen. *Bulbochaete* hair cell, × 29 000. (Micrograph courtesy of T. W. Fraser).

Microfilaments

Fine strands, which have become known as microfilaments, are being discovered in more and more cells, plant and animal. It is likely that they participate in the movement of cytoplasm and its components (see also Plate 27b); thus while microtubules may be likened to intracellular bones, microfilaments may be regarded as some form of intracellular muscle.

Plate 43g - i These phase contrast photomicrographs were taken at approximately 30 second intervals, and illustrate the same part of a living hair on a petiole of *Heracleum mantegazzianum*. One fibrous strand (microfilament bundle) is in focus in all three photomicrographs, but others are also visible. Cytoplasmic streaming is particularly vigorous near the fibres. Comparison of the three micrographs shows that a plastid (P) moved diagonally across the field, while there are much more dramatic changes in the population of mitochondria (M). × 3300; (micrographs courtesy of Drs. T. P. O'Brien and M. E. McCully, reproduced by permission of Springer-Verlag from *Planta*, **94**, 91 (1970)).

Plate 43j It is difficult to detect microfilaments by electron microscopy unless, as here, appreciable lengths lie in the plane of the ultra-thin section. In this example the microfilaments, each 2-5 nm in diameter, mostly lie parallel to one another, but the bundles frequently change direction. Spinach root tip cell, × 47 000.

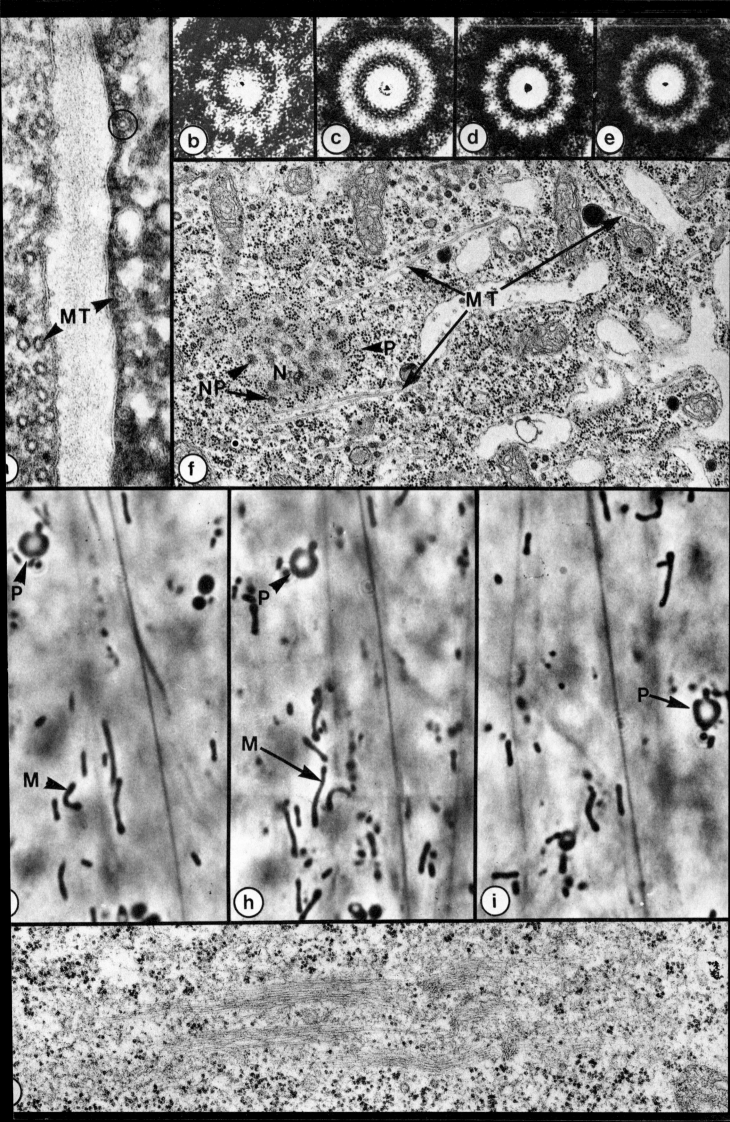

Plate 44

Cell Division (1): Mitosis in *Haemanthus*

This set of photomicrographs was taken using the Nomarski interference-contrast technique. The sequence shows the course of mitosis in a single living endosperm cell from the blood lily, *Haemanthus katherinae* Bak. These cells can be removed from young fruits, and, because they have no cell wall, they can be spread on a microscope slide and sequential observations made as they undergo division. The cumulative total time from the first micrograph in the sequence (a) is recorded at the end of each caption below. All the micrographs are at the same magnification (× 700). The identification of some of the smaller objects is based on electron microscope studies of the same material.

Plate 44a - c *Prophase.* The nuclear envelope (NE) forms a distinct boundary between the nuclear contents (condensing chromosomes, CH; disintegrating nucleolus, NL) and the developing clear zone surrounding the nucleus (see also nucleus N-1 in Plate 1). In this case the clear zone is 3-polar (asterisks in (b)). The small protrusions of the nuclear envelope at the poles are where microtubules from different directions intermingle. As prophase progresses, the 3-polar condition gives way to the normal bi-polar division figure (c). The cytoplasmic components are arranged around the periphery of the cell: the numerous small vacuoles (V) are quite conspicuous, but others are not readily identifiable, except for a large mitochondrion (M) which can be seen in (a). Times: (a) 0; (b) 15 min; (c) 22 min.

Plate 44d - f *Prometaphase-metaphase.* Following rupture of the nuclear envelope (between (c) and (d)), a normal bi-polar spindle (poles at asterisks in (d)) has now been formed. The spindle fibres are as yet barely discernible. Movements of chromosomes during prometaphase (d) and (e) result in the orientation of their kinetochores (points of attachment to spindle fibres) at the equator (E-E) by the time of metaphase (f). Kinetochore fibres (arrowheads) are particularly prominent in the upper half spindle. That the chromosomes are double, consisting of chromatids twisted around one another, has been visible since early prophase (a): the sister chromatids now gradually untwist ((d) and (e)), prior to their separation in the next stage of mitosis. Times: (d) 1 h. 2 min; (e) 1 h 18 min; (f) 1 h 40 min.

Plate 44g - i *Anaphase.* Movement of the sister kinetochores to opposite poles gives the typical trailing arm chromosome configurations of early (g) and mid (h) anaphase. At these stages the kinetochore fibres (arrowheads) are still visible. By late anaphase (i) the chromosomes have begun to coalesce at the poles. Phragmoplast fibres now develop at the centre of the spindle, and their activity results in lateral movements of the trailing chromosome arms: comparison of (i) and (h) shows how they fan out from their previous alignment along the axis of the spindle. Dictyosomes (D) begin to invade the phragmoplast region. Times: (g) 1 h 50 min; (h) 1 h 56 min; (i) 2 h 7 min. (Notice the rapidity of the movements of chromosomes during anaphase, as compared with during prometaphase and metaphase).

Plate 44j - l *Telophase – early cytokinesis.* Progressive condensation of the two chromosome masses results in the formation of two daughter nuclei. Each becomes bounded by its own nuclear envelope (NE in (l)). A comparable stage is shown in nucleus N-2 of Plate 1, where a newly-regenerating nucleolus lies amongst the uncoiling chromosomes. Between the two nuclei the phragmoplast fibres develop further (F in (j)) and material accumulates along the former equator, giving rise to the cell plate (CP in (k) and (l)). Dictyosomes (D) in side view and face view ((k) and (l) respectively) are prominent among the fibres. A very long and attenuated mitochondrion (M) lies across the cell plate in (l). Had the cells not been removed from the fruit, the cell plates formed at the division illustrated in this plate would have developed further to become the first cell walls of the previously naked endosperm tissue. Times: (j) 2 h 13 min; (k) 2 h 22 min; (l) 2 h 40 min.

(Micrographs provided by Dr. A. S. Bajer).

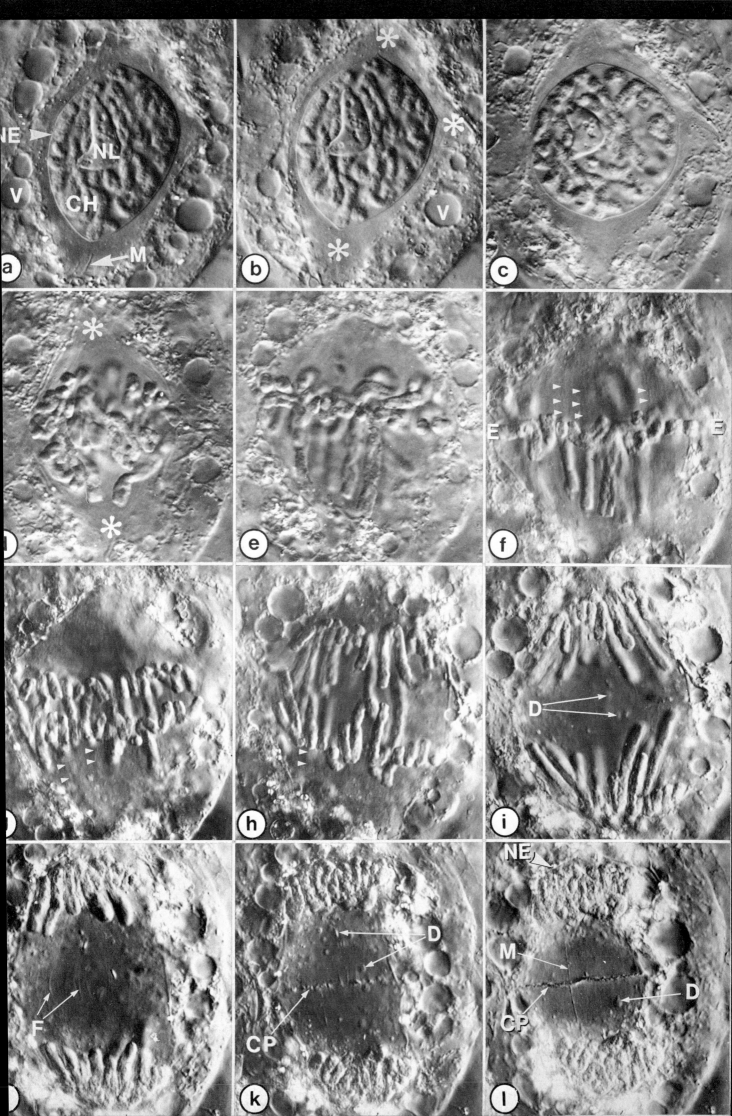

Plate 45

Plates 45-48 display electron micrographs of ultra-thin sections, complementary to the light micrographs of Plate 44. Unless stated otherwise, all of the sections are longitudinal with respect to the cell axis. Where possible, the micrographs are oriented so that the spindle poles lie at the top and bottom of each figure.

Cell Division (2): Prophase

Plate 45a Prophase nuclei (corresponding to Plate 44 a-c) show condensed chromatin (CH) and the initial stages of dispersal of nucleolar material (NL). Segregation of granular and fibrillar zones is evident in the nucleolus. The nucleoplasm, still enclosed by the intact nuclear envelope (NE), shows little structure at this magnification. There is no obvious clear zone around the nucleus, indeed proplastids (P), mitochondria (M) and vacuoles (V) are all lying close to the nuclear envelope. × 6500.

Plate 45b The area enclosed by the rectangle in (a) is shown at higher magnification in (b). NL marks a mass of nucleolar fibrils, and nucleolar granules are at upper left. Part of a chromosome (CH) is also included. The background nucleoplasm is relatively empty at this stage (cf. below). × 28 000.

Plate 45c - e The areas enclosed by the rectangles in (c) are shown enlarged in (d) and (e). At this stage (approximately between (d) and (e) of Plate 44) the nuclear envelope (NE), with its pores (NP in (e)), is disintegrating and a number of breaks have occurred (see asterisks in (c) and (e)). Although the nuclear envelope has only been partially removed, the nucleoplasm between the chromosomes (CH) has been invaded by a large number of ribosomes, mostly in the form of polyribosomes (compare (c) and (d) with (a) and (b) above). Vesicles (VE) also appear in the former nucleoplasm (note the presence of dictyosomes (D in (c)) outside the disintegrating envelope). Components of the future spindle have made their first appearance in the nucleoplasm: microtubules (MT in (d) and (e)) are now evident, near breaks in the nuclear envelope (e), as well as in the outer region of the nucleoplasm (e) and in central regions (d). The nucleolus (NL in (c) and (d)) appears more dispersed than in (a).

One of the most striking changes is in the nature of the nuclear envelope. When intact, it bears pores, and ribosomes are bound only to the cytoplasmic face of the outer membrane (Plates 18d, 43f). It has been caught here in a state in which it more closely resembles cisternae of rough endoplasmic reticulum. Thus it has been ruptured, and ribosomes are bound to both the inner (solid arrows in (e)) and the outer membrane. Pores are still present. A similar state recurs at telophase of mitosis (Plate 47d).
(c) × 16 000; (d) and (e) × 50 000.
(a) - (e) all from root tip cells of *Vicia faba*.

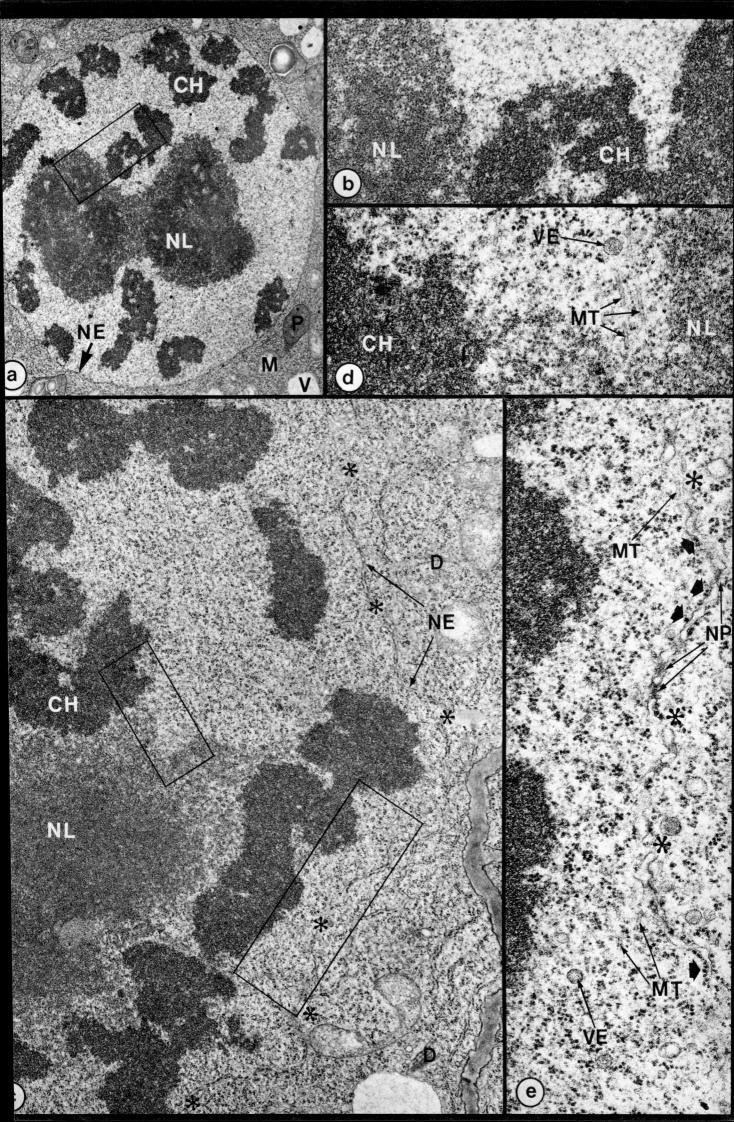

Cell Division (3): Prometaphase and Metaphase

Plate 46a This micrograph continues the series started in Plate 45a and c. It is equivalent in stage to Plate 44d, that is, prometaphase, and shows a small sector of the clump of chromosomes liberated from a nucleus. Microtubules, which make their appearance in the nucleoplasm at the time of the rupture of the nuclear envelope (Plate 45d, e), are now present in extensive arrays (MT). The chromosomes (CH) have not yet become aligned on the equator of the division figure. The nucleolus (NL) is so dispersed as to be scarcely detectable. *Vicia faba* root tip, × 11 000.

Plate 46b, c and d illustrate stages of mitosis in cells in root tips of white lupin (*Lupinus albus*). The most obvious difference between lupin and the broad bean (*V. faba*) used in Plates 45 and 46a is that the chromosomes (and the cells) are very much smaller in the former.

Plate 46b As was the case in (a), the chromosomes shown here were fixed during prometaphase movements, and hence lie at different, non-equatorial, levels of the division figure. The microtubule (MT) system has developed sufficiently to be called the spindle. The microtubules are in bundles, and are oriented longitudinally, in the pole-to-pole axis. Those seen in this micrograph do not, however, seem to pass from pole to pole. Rather they extend longitudinally from the chromosomes (CH). They are, in other words, the microtubules that constitute the kinetochore fibres of Plate 44f. × 36 000.

Plate 46c This off-centre longitudinal section of a metaphase cell (equivalent to Plate 44f) shows the paired chromatids (CH) lying near the periphery of the equatorial region of the spindle. Kinetochore fibres (KF, hardly visible at this magnification) run from the kinetochores, oppositely oriented on the chromatid pairs, and mix with other fibres in the spindle, all passing towards the poles. The cell components (proplastids (P), mitochondria (M), dictyosomes (D), lipid droplets (L) and vacuoles (V)) are in general excluded from the spindle region and lie between it and the cell wall (just outside this micrograph). × 8000.

Plate 46d Kinetochores vary in structure throughout the eukaryotes. Here the electron microscope reveals little but a matrix (K) in which the kinetochore microtubules (KM) terminate. Other microtubules (MT) penetrate between the chromosomes (CH), and probably are examples of pole-to-pole microtubules (as distinct from pole-to-kinetochore).

The inserts present the alternative view of a kinetochore, that is, as seen in sections cut in the plane at right angles to that of the main micrograph, and at a position equivalent to the level of the arrows from the letters K. The material was a dividing *Chlorella* cell. The upper insert shows transversely sectioned microtubules (ringed), one single, and one pair, in each case surrounded by fuzzy material. In the *adjacent* section (lower insert) of precisely the same area, the microtubules are no longer visible, and more of the fuzzy kinetochore or chromosomal material is included. The microtubules clearly are of the kinetochore type, and the two sections must have spanned the microtubule termini in or on the chromatids.
× 36 000; inserts × 45 000.
(Inserts provided by Dr. A. W. Atkinson, Jr.)

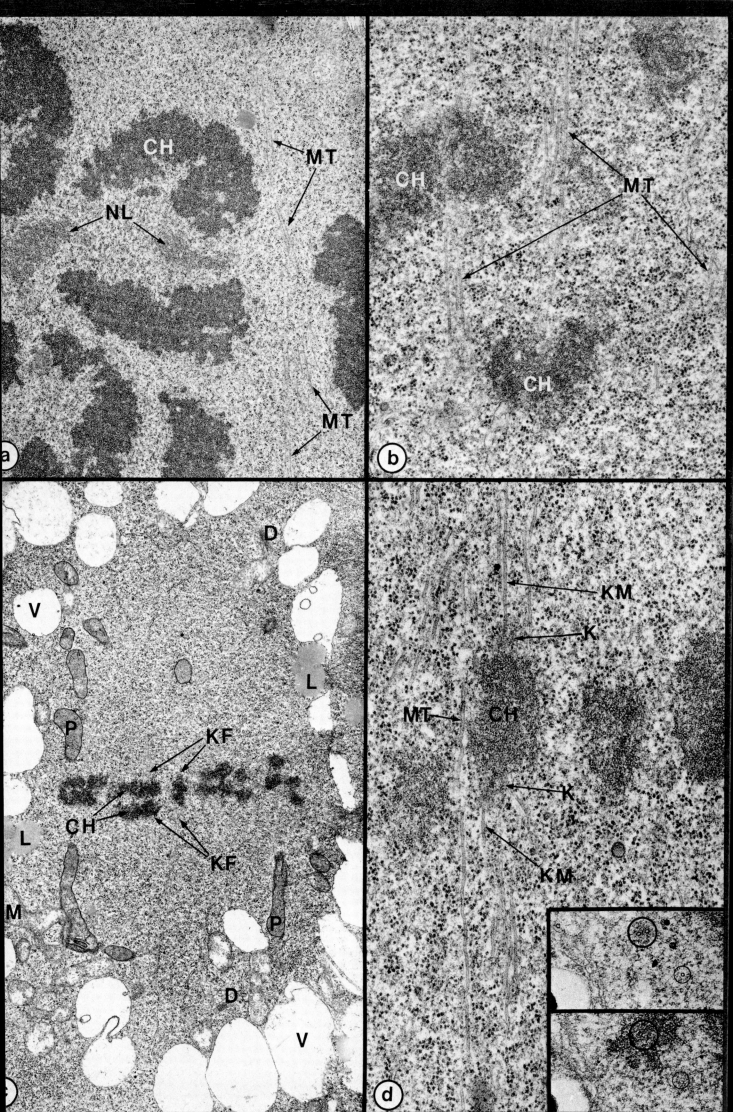

Plate 47

Cell Division (4): Anaphase—Early Telophase

Plate 47a This mid-anaphase cell (equivalent to Plate 44 g-h) from a white lupin root tip shows parts of several chromatid arms, caught within the section at various positions between the equator (which runs approximately left to right) and the spindle poles (at top and bottom). Two sister chromatids, each carrying a nucleolar organizer region (NO), lie close to each other at the left hand side of the spindle. The nucleolar organizer is clearly different from the chromatid material on either side of it; light microscope staining reactions indicate that the concentration of DNA is low in the organizers. The whole spindle region, filled with numerous ribosomes and portions of sectioned spindle microtubules (MT), is surrounded by many cisternae of rough endoplasmic reticulum (ER), outside which lie other components of the cytoplasm: dictyosomes (D), mitochondria (M), plastids (P), and vacuoles (V). × 8000.

Plate 47b and c Later in anaphase, when the chromosomes begin to coalesce at the polar regions (upper parts of both (b) and (c); stage equivalent to Plate 44i), each chromosome becomes surrounded by a zone of granular material (large arrows in (b)) which is separated from the chromatin by a narrow electron-transparent space. Micrograph (b) shows a trailing chromosome arm possessing a nucleolar organizer region (NO), and a further example of the same structure is included in (c). Microtubules (MT), seemingly attached to the chromosomes, are seen in (b). Note also the vesicles (VE) amongst the trailing arms. Elements of endoplasmic reticulum modified by the development of nuclear pores (NP) are also present close to the chromosomes. *Vicia faba* root tip, (b) × 16 000; (c) × 10 000.

Plate 47d By early telophase (equivalent to Plate 44 j-k) a nuclear envelope with pores (NP) invests most of the coalescing chromosomes. The insert shows the new nuclear envelope in more detail, including a pore (large arrowhead), and, where it is not closely appressed to chromosome material, ribosomes on *both* surfaces. Some of the ribosomes on the future inner membrane are arrowed. At this stage the nuclear envelope therefore resembles the fragments present soon after the end of prophase (Plate 45e), but it is not clear whether such fragments persist throughout metaphase and anaphase to contribute to the re-forming envelope at telophase, or whether the prophase fragments lose their pores to become typical rough endoplasmic reticulum, with some cisternae later on re-synthesizing pores and metamorphosing back to the nuclear envelope condition.

The nucleolar organizer region (NO) has started to expand at its peripheral regions, differentiating an outer granular layer (G). Further expansion leads to regeneration of the nucleolus, as in nucleus N-2 in Plate 1, which is from the same material as the present micrograph (*Vicia faba* root tip), but is at a slightly later stage of telophase, with a larger nucleolus and with the chromosomal material beginning to de-condense towards the interphase condition.
× 32 000, insert × 52 000.

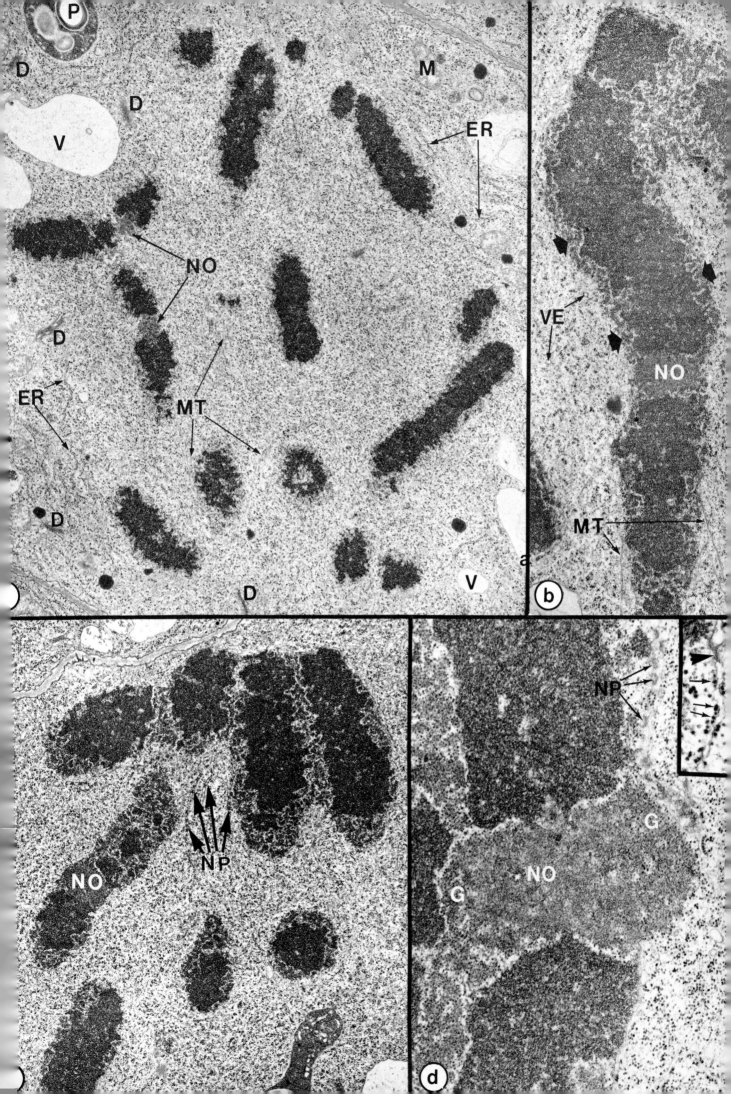

Plate 48

Cell Division (5): Telophase and Cytokinesis

Plate 48a This off-centre longitudinal section of a *Beta vulgaris* root tip cell is equivalent in stage to Plate 44 k-l. It shows two daughter telophase nuclei (N), each bounded by a nuclear envelope (NE) that is by now complete. Between the two lie phragmoplast microtubules (MT) and the developing cell plate (CP). The discrete vesicles (VE) of the young cell plate have in places begun to coalesce to form larger units (large arrow). Continuation of coalescence gives rise to the plasma membranes of the cross wall separating the daughter cells. A dictyosome (D) is present beside one of the nuclei, but is inconspicuous: none is seen (in this section) in the vicinity of the developing cell plate. × 13 500.

Plate 48b Details of phragmoplast microtubules and a young cell plate are seen here at a similar stage to that in (a). Long microtubules (MT) interdigitate (large arrows) at the layer of vesicles (VE) which form the cell plate. Vesicles are also present throughout the cytoplasm on both sides of the cell plate (small arrows). *Vicia faba* root tip cell, × 23 000.

Plate 48c It is very difficult to obtain face views of the cell plate—that is, in sections cut at right angles to that of (a). Such a section would be seen edge-on in the view shown in (a), and would be only a fraction of a millimetre in thickness at the magnification of (a). A face view of part of a cell plate at a late stage of its development is presented in (c). Due to its undulating contour, the plate passes into and out of the section, so that part of the micrograph includes coalescing vesicles of the plate, and part neighbouring cytoplasm. Some vesicles (large arrows) remain as discrete spheres between the advancing cell plate and the side wall (CW). About 150 profiles of cross-sectioned phragmoplast microtubules (arrowheads) are seen throughout the micrograph. They are most concentrated amongst the coalescing vesicles and tubules of the cell plate. *Avena sativa* anther, × 47 000.

Plate 48d, e, f Small segments of cell plates are shown here in side profile (as in (a)) in order to demonstrate certain points of detail. Regions of the coalescing membrane surface (black arrows in (d), and just to the left of ER in (f)) are coated with an array of fuzzy spikes, forming a layer some 15 nm in thickness on the cytoplasmic face of the membrane. Such surfaces resemble the fuzzy coating of coated vesicles (Plate 27a). It is not possible to say from a static image such as this one whether coated vesicles gave rise to the coated areas of the cell plate by fusing with it, or whether coated vesicles are being formed at the cell plate, and have been caught just prior to their release into the cytoplasm. A similar dilemma arose in connection with Plate 28a. Alternatively the surface coat might have properties and functions independent of vesicle arrival or formation.

A second point, illustrated in (d) and (f), is that cisternae of endoplasmic reticulum (ER) can become trapped in a position such that they interconnect the daughter cells that are being separated by the growing cell plate. This could be the source of the axial structures of plasmodesmata (PD, see also Plate 14a, c, d), thought to be derivatives of endoplasmic reticulum.

A point of contact between the side wall (CW) of the dividing parental cell and the extending cell plate is shown in (e). The dark objects in the parent cell wall are remains of plasmodesmata, which look as if they may have become occluded. As in (d) and (f), the coalesced vesicles of the cell plate contain fibrillar material which is the first sign of the primary wall that will separate the two daughter cells. Similar material is seen in the protuberance which is emerging from the parent wall towards the cell plate. Nearby in the cytoplasm on both sides of the plate are numerous vesicles (VE), amongst the phragmoplast microtubules (MT). Some of the latter may be terminating at the cell plate, but as the section is at an angle relative to the plane in which most of the microtubules lie, this cannot be detected with certainty. Here and there hemispherical profiles on the membrane of the cell plate (open arrows in (d) and (e)) suggest that vesicles such as those lying free (VE) have fused with and delivered their contents to the plate (but see above comments on coated vesicles).

Vicia faba root tip cell, (d) × 67 000, (e) × 55 000, (f) × 55 000.

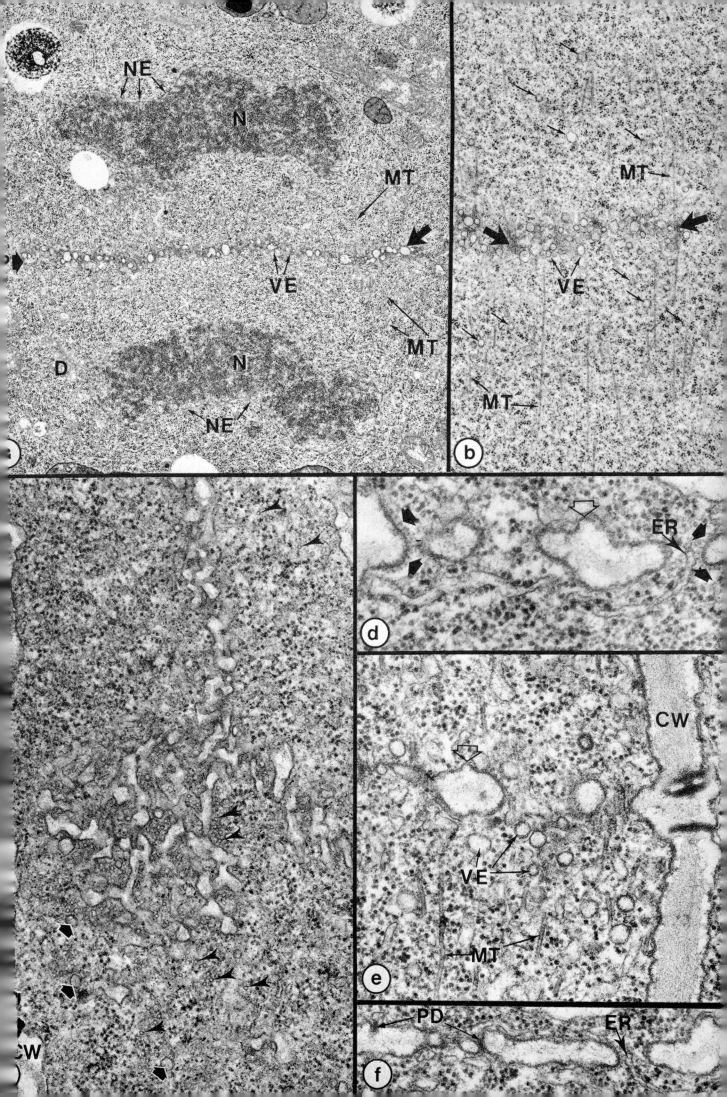

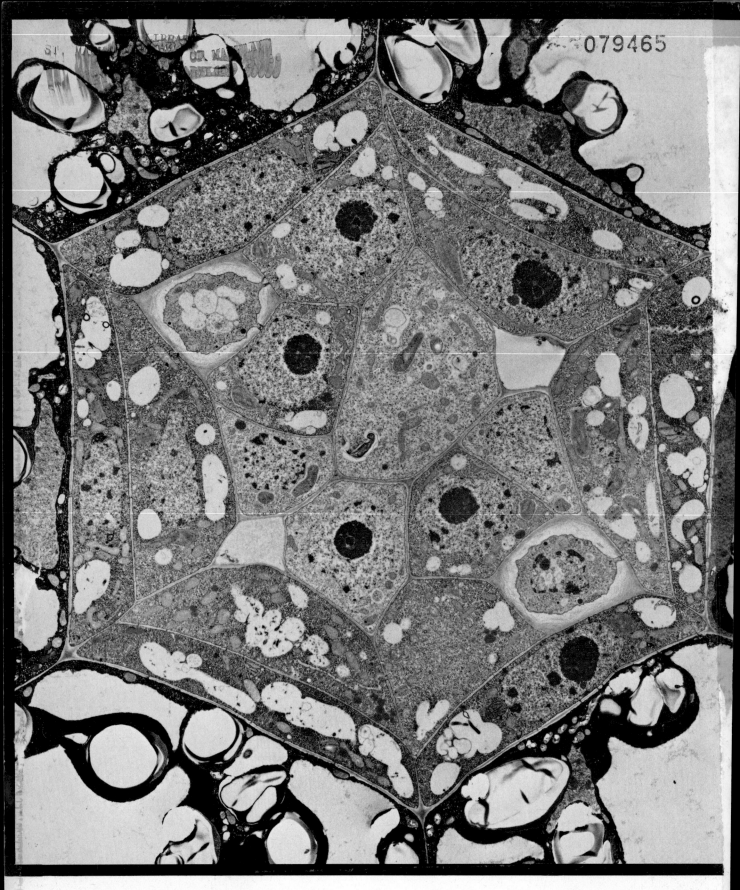

Plate 49 Plates 1–48 have illustrated some details of individual cells and components of cells. The purpose of the final plate is different, and largely symbolic: to emphasize by its precision and symmetry that there is another aspect of plant cell biology—the structural and functional integration of cells in the tissues and organs of the plant. The cells that are seen here are juvenile, in a cross section cut near the tip of the highly miniaturized root of the water fern *Azolla* (× 5800). Despite the compact nature of the overall structure, and their close proximity, the 22 cells specialize along 6 different pathways of maturation. The mature state is shown, with labels, in Plate 16a. When mature, the cells are disposed in unchanged numbers and in an unchanged geometrical relationship, in which the 6 cell types collaborate to perform the multifarious functions of the vascular system of the root. Every cell type plays its specialized part: no one could operate on its own. Just as specialized sub-cellular components participate in a collaborative group existence upon which depends the survival of the cell as an entity, so it is with the cells of the plant. Their collaboration creates functional entities at levels of organization higher than that of the cell—and beyond the scope of the present book.